The world is currently experiencing unprecedented global change, with population increase, urbanisation, climate change and environmental degradation combining to make management of freshwater resources a critical policy focus of the twenty-first century. This timely book designs and develops an original, analytical framework for water law reform processes, using case studies across four jurisdictions.

Addressing the four principal areas of water law – integrated water resource management (IWRM) and river basin planning, water rights and allocation, water pollution and quality, and water services – this book provides a comprehensive study of water law, within the context of global and regional policy agendas. Case studies from England, Scotland, South Africa and Queensland, Australia, are presented, providing comparators from both common law and mixed jurisdictions, from the northern and southern hemispheres, and from developed and developing countries. A legislative framework is proposed for water law reform processes, and the consequences of different reform options are considered and investigated.

A valuable resource for academics and graduate students in environmental law, resource management, hydrology and social science, this book is also highly relevant to policymakers, NGOs and legal practitioners.

SARAH HENDRY is a Lecturer in Law at the University of Dundee in the Centre for Water Law, Policy and Science, a Category 2 Centre under the auspices of UNESCO, where she specialises in comparative national water law, with special expertise in EU and Scots water and environmental law. She is Director of the Centre's taught programmes and Advisor of Studies for its research students, as well as guest lecturer in water and environmental law at UNESCO IHE in Delft, the Netherlands. Her interests extend across both water resources and water services, and she sits on the Scottish Customer Forum, a new body to better represent consumers in price setting for water services.

INTERNATIONAL HYDROLOGY SERIES

The **International Hydrological Programme** (IHP) was established by the United Nations Educational, Scientific and Cultural Organization (UNESCO) in 1975 as the successor to the International Hydrological Decade. The long-term goal of the IHP is to advance our understanding of processes occurring in the water cycle and to integrate this knowledge into water resources management. The IHP is the only UN science and educational programme in the field of water resources, and one of its outputs has been a steady stream of technical and information documents aimed at water specialists and decision-makers.

The **International Hydrology Series** has been developed by the IHP in collaboration with Cambridge University Press as a major collection of research monographs, synthesis volumes, and graduate texts on the subject of water. Authoritative and international in scope, the various books within the series all contribute to the aims of the IHP in improving scientific and technical knowledge of freshwater processes, in providing research know-how and in stimulating the responsible management of water resources.

Frameworks for Water Law Reform

Sarah Hendry

Centre for Water Law, Policy and Science, University of Dundee

CAMBRIDGE
UNIVERSITY PRESS

CAMBRIDGE
UNIVERSITY PRESS

University Printing House, Cambridge CB2 8BS, United Kingdom

One Liberty Plaza, 20th Floor, New York, NY 10006, USA

477 Williamstown Road, Port Melbourne, VIC 3207, Australia

4843/24, 2nd Floor, Ansari Road, Daryaganj, Delhi - 110002, India

79 Anson Road, #06-04/06, Singapore 079906

Cambridge University Press is part of the University of Cambridge.

It furthers the University's mission by disseminating knowledge in the pursuit of
education, learning and research at the highest international levels of excellence.

www.cambridge.org
Information on this title: www.cambridge.org/9781108446730

First published 2015
First paperback edition 2017

A catalogue record for this publication is available from the British Library

Library of Congress Cataloging in Publication data
Hendry, Sarah, 1963-
Frameworks for water law reform / Sarah Hendry, UNESCO Centre for Water Law,
Policy, and Science, University of Dundee.
        pages cm. – (International hydrology series)
Includes bibliographical references and index.
ISBN 978-1-107-01230-1
1. Water–Law and legislation.   2. Water resources development–Law and legislation.
3. Water rights (International law)   4. Water-supply–Law and legislation.   5. Water
transfer–Government policy.   6. Integrated water development.   7. Law reform.
I. Title.
K3496.H46 2015
346.04´691–dc23   2014022147

ISBN 978-1-107-01230-1 Hardback
ISBN 978-1-108-44673-0 Paperback

This book is dedicated to my Mother, and to Professer Mike Bonell.

Contents

Acknowledgements

I owe grateful thanks to many people, without whom this book would not exist. When I began my PhD studies, which formed the original research, in 2001, I would not have envisaged where the journey would lead. From that period, I would especially thank my supervisors at the University of Dundee, Patricia Wouters and Colin Reid, and the head of the PhD programme, Chris Rogers. Also my then colleagues at the University of Abertay: the lawyers, especially Jim Murphie, Ken Swinton, Jim Tunney, Maria O'Neill and Fiona Grant; Nicholas Terry and Mary Malcolm; and in the Urban Water Centre, Richard Ashley, Chris Jefferies and David Blackwood. Without them, I would not be working in water.

After completion of the thesis, it was my privilege to move to the Centre for Water Law, Policy and Science (a Category 2 Centre under the auspices of UNESCO) at the University of Dundee; and, further, to have the original thesis accepted for publication in the Cambridge University Press International Hydrology Series in association with UNESCO. It took several years, and several setbacks, to produce the final book. For their patience and help I would like to thank all the people at Cambridge University Press, especially Susan Francis, Rosina Piovani and Zoe Pruce.

At the Water Centre here, as well as Pat Wouters, my thanks go to Andrew Allan and Alistair Rieu-Clarke, Chris Spray, Geoff Gooch, and especially to Michael Bonell; also to Ian Ball, Janet Liao, Janeth Warden-Fernandez and the academic and administrative staff in the Graduate School. In the Law School, Colin Reid and Andrea Ross; and in Geography, Alan Werritty and Alison Reeves.

In the world of water, I have learnt so much from so many. It is not possible to name them all, but much gratitude to those in Australia and South Africa who patiently answered my questions and offered me hospitality; especially Poh Lin Tan, Mark Pascoe, Brian MacIntosh, Sarah Goater, Alex Gardner, Jennifer McKay, Marius Claasen, Kate Tissington and Robyn Stein.

I would also like to thank my students, both at Abertay and at Dundee. There is always much to learn from them and I wish them every success in their endeavours.

Most of all, my thanks go to my family and friends. Again, too many to name, but especially Mark and Emma, Scott and Georgia, Duncan and Stephanie, Maggie, Susan, Judith, Cornelia and Louise, Jim and Gill, and the children large and small – Zander, Jamie, Nicki, William and Grace. Thank you for a world outside water and the law.

In the several years that elapsed between completing the thesis and completing the book, many things changed; not least, the law in each jurisdiction, making it a daunting task. I hope that I have been able to incorporate much that I have learnt – about water, as well as water law – to make the final product much better than it would have been without the intervening years. The law is stated as at 1 February 2014; whilst many of the people above have contributed their time and thoughts, any errors or misconceptions of course remain my own.

Abbreviations

AMCOW	African Ministerial Council on Water	DNRM	Department of Natural Resources and Mines (Queensland)
ANZECC	Australian and New Zealand Environment and Conservation Council	DOL	Distribution Operations Licence
APSC	Australian Public Service Commission	DoRA	Distribution of Revenue Act (South Africa)
ARMCANZ	Agriculture and Resource Management Council of Australia and New Zealand	DPMAG	Diffuse Pollution Management Advisory Group
AUD	Australian dollars	DWA	Department of Water Affairs
BOO	build–own–operate	DWAF	Department of Water Affairs and Forestry
BOOT	build–own–operate–transfer	DWQMP	Drinking Water Quality Management Plan
CAMS	Catchment Abstraction Management Strategies	EA	Environment Agency
		ECHR	European Convention on Human Rights
CAP	Common Agricultural Policy	EIA	environmental impact assessment
CAR	Water Environment (Controlled Activities) (Scotland) Regulations	ELL	cconomic level of leakage
		EP	Environmental Protection Regulation (Queensland)
CBD	Convention on Biological Diversity		
CMA	Catchment Management Agency	EPA	Environmental Protection Act (Queensland)
CMA	Competition and Markets Authority	EPBC Act	Environmental Protection and Biodiversity Conservation Act (Australia)
COAG	Council of Australian Governments		
CoGTA	Department of Cooperative Governance and Traditional Affairs	EPR	Environmental Permitting Regulations (England and Wales)
COHRE	Centre on Housing Rights and Evictions	EQS	environmental quality standards
CSS	customer service standard	ERA	environmentally relevant activity
DEA	Department of Environmental Affairs (South Africa)	EU	European Union
		FAO	Food and Agriculture Organization of the United Nations
DEAT	Department of Environmental Affairs and Tourism (South Africa)		
		FBSan	Free Basic Sanitation
DEFRA	Department for Environment, Food and Rural Affairs (UK)	FBW	Free Basic Water
		GBR	general binding rule
DEHP	Department of Environment and Heritage Protection (Queensland)	GC15	General Comment 15
		GL	gigalitre
DERM	Department of Environment and Resource Management (Queensland)	GWP	Global Water Partnership
		HC	House of Commons
DETR	Department of the Environment, Transport and the Regions (UK)	HL	House of Lords
		HRC	Human Rights Council
DEWS	Department of Energy and Water Supply (Queensland)	ICESCR	International Convention on Economic, Social and Cultural Rights
DHS	Department of Human Settlements (South Africa)	IWRM	integrated water resource management
		LPD	litres per person per day

MDGs	Millennium Development Goals	SEQ Water Act	South East Queensland Water (Distribution and Retail Restructuring) Act
ML	megalitre		
MSA	Municipal Systems Act (South Africa)	Standards	Regulations Relating to Compulsory National
NEMA	National Environmental Management Act (South Africa)	Regulations	Standards and Measures to Conserve Water
		SUDS	Sustainable Urban Drainage Systems
NGO	non-governmental organisation	SW	Scottish Water
NHMRC	National Health and Medical Research Council	Tariffs Regulations	Norms and Tariffs in Respect of Water Services Regulation (South Africa)
NRM	Natural Resource Management	UKTAG	UK Technical Advisory Group
NRMMC	National Resource Management Ministerial Council	UN	United Nations
		UNDESA	United Nations Department of Economic and Social Affairs
NWA	National Water Act (South Africa)		
NWC	National Water Commission	UNDP	United Nations Development Programme
NWI	National Water Initiative	UNEP	United Nations Environment Programme
NWQMS	National Water Quality Management Strategy	UNICEF	United Nations Children's Fund
NWRS	National Water Resource Strategy	USD	US dollars
OECD	Organisation for Economic Co-operation and Development	UWWTD	Urban Waste Water Treatment Directive
		WA2003	Water Act 2003
OFWAT	Office for Water Services (Water Services Regulation Authority)	Water Policy	Environmental Protection (Water) Policy
		Water Supply Act	Water Supply (Safety and Reliability) Act (Queensland)
PPIAF	Public–Private Infrastructure Advisory Forum		
PPP	public–private partnership	WEWS	Water Environment and Water Services Act (Scotland)
PSP	private sector participation		
QCA	Queensland Competition Authority	WFD	Water Framework Directive
QWA	Water Act (Queensland)	WHO	World Health Organization
RBD	river basin district	WIA	Water Industry Act (England)
RBMP	River Basin Management Plan	WICS	Water Industry Commission for Scotland
ROL	Resource Operations Licence	WISA	Water Industry (Scotland) Act
ROP	Resource Operations Plan	WQM	Water Quality Management
RPA	Rural Payments Agency	WRA	Water Resources Act (England)
RQO	Resource Quality Objective	WRC	Water Research Commission
SADC	Southern African Development Community	WRP	Water Resource Plan
SAMP	Strategic Asset Management Plan	WSA	Water Services Act (South Africa)
SANS	South African National Standards	WSP	water services provider
SEA	Strategic Environmental Assessment	WUA	Water Users' Association
SEPA	Scottish Environment Protection Agency	WWF	World Wide Fund for Nature (World Wildlife Fund)
SEQ	South East Queensland		

1   Policy context

Water is the basis of all things.

Thales (640 BC)

1.1   INTRODUCTION

Water is the stuff of life – 70% of the planet,[1] 60% of the human body.[2] Its symbolic and cultural aspects are represented in religion,[3] philosophy,[4] and every branch of the arts. It is a topic of academic study – in the hard sciences, by hydrologists and geologists, biologists and chemists, geographers and engineers, and of course ecologists, not to mention the medical professions. In the social sciences, it is studied by economists, sociologists, political scientists – and lawyers. As the precious resource is put under increasing pressures, more and more professionals are engaged; too often, the role of lawyers is overlooked save in the negative – 'we'll have to bring the lawyers in' – almost a threat, and equally unwelcome to every party round any table. Yet the law creates the framework and the ground rules within which the resource is managed, and provides the mechanisms by which subsequent disputes are resolved. Logically, the better the provision for the former, the fewer occasions arise for the latter.

The title of this book is *Frameworks for Water Law Reform* and the aim is to consider what provision should be made when reforming a national water law. The ambit will include the principal elements of water resource management, including water allocation and water quality, and also water services (defined here as the supply of drinking water and sanitation services). It will make a comparative analysis of four jurisdictions where water law is currently being, or has recently been, reformed – England, Scotland, South Africa and Queensland,

Australia. It does not offer a 'model', in the sense of a single or best solution. Rather, it sets out a framework, identifying the key elements of a modern water law and the various ways in which these could be established. It is hoped the results will be useful to those engaging in or considering a reform process, not just to states or public agencies, but also to other parties, including non-governmental organisations (NGOs), as well as water professionals, students and, of course, lawyers in academia and in practice.

1.1.1   Human and social issues

To say there is a world water crisis is trite, yet it may bear repetition. The 'headline' figures are well known – still nearly 800 million people without access to improved drinking water supplies, still around 2.5 billion without improved sanitation.[5] Of the top five communicable diseases worldwide, two – diarrhoea and malaria – are directly linked to water, and all are affected by the lack of sufficient water and, especially, sanitation.[6] An estimated 10% of the total global burden of disease, and 6.3% of all deaths, could be prevented by access to improved water, sanitation and hygiene.[7] There is a disproportionate effect on young children and the elderly, on women (in terms of maternal health and the burden of caring for the sick), regionally in sub-Saharan Africa and Southern Asia, and globally for those living in extreme poverty. Better provision directly affects social and economic wellbeing, enabling more time to be spent on productive activities and more girl children to attend school. Meantime the global population is increasing,[8] and so are the pressures on the resource. Approximately one-third of the world's population lives in countries that are water stressed, and this is predicted to increase to as much as two-thirds by 2025.[9] Water is a cross-cutting issue: it affects public and individual health; it is a critical resource for primary and secondary production; it impacts directly and indirectly on economic and social wellbeing; and it disproportionately affects the poor and dispossessed.

---

[1] Pidwirny (2006).

[2] Although the figure is variable, dependent on age, gender and levels of fat; see 'MadSci Network' http://www.madsci.org/posts/archives/may2000/958588306.An.r.html.

[3] Every creation myth begins with the emergence of life from some great ocean, physical or metaphysical; see, for discussion of the universality, Ball (2002).

[4] The early philosophers studied the natural world; Thales, 640 BC, wrote that 'water is the basis of all things', see 'Ancient Greek Philosophy' http://www.iep.utm.edu/g/greekphi.htm. For an alternative perspective on the abstract and spiritual nature of water, see Emoto (2004).

[5] WHO/UNICEF (2012). The statistics, and the terminology, will be discussed in Chapter 5.

[6] UN-Water (2006).   [7] Pruss-Ustun *et al.* (2008).

[8] Currently around 7.2 billion, and predicted to rise to 9.6 billion by 2050; UNDESA (2012).

[9] UN-Water (2009).

1.1.2   Environmental issues

Whilst 70% of the world's surface is covered in water, only 2.5% of that is freshwater, and nearly 70% of that is locked in the Arctic and Antarctic.[10] Of the remainder, some 30% is groundwater, permafrost or swamp water; these sources include 97% of water available for human use. Surface waters (rivers and lakes) amount to just 0.3% of global freshwater, and the total available freshwater supply for humans and ecosystems is less than 1% of the whole freshwater resource, and 0.01% of all global waters. Whilst the freshwater cycle is theoretically self-cleansing and renewing, as pollutants enter the cycle and the resource is over-exploited it becomes more difficult to sustain this natural process. Meantime climate change affects the water cycle and water availability in numerous ways, not all of them predictable, but likely to include more extreme weather events, including storm, flood and drought, and the melting of the glaciers. The net effects will be felt not just by human populations, but by all the interconnected ecological systems on which life depends. Water can be a source, a pathway and a receptor; but for humans it is also a driver of change. Populations must move to find water, societies cannot develop without water, it is non-substitutable, and without it there is no life as we know it.

1.1.3   Why water law?

Given the scale of the problems, one might ask how law could play more than a bit part. Law gives the structure within which other actors play their roles; it provides mechanisms for decision-making, participation and conflict resolution. Because it sets the structure, once in place, other socio-economic and political activities work within that legal environment, and actors in those realms generally consider 'the law' to be a set of unchangeable factors, at least in the short to medium term. It is important to get the framework right.

Water law operates at different levels: international, transnational and national. International law concerns the relations between states, in the form of treaties or conventions, as well as customary international law. Transnational law is a term used to address the convergence of laws in a globalising world, especially in world trade, but also international investment, including some aspects of water services law. National law operates within states, or at sub-state level, such as local laws, as well as customary law, and it is national water law with which this work is concerned. All over the globe, as states reform their water management provision in line with global policies, they also review their national laws. However, although there are extensive academic writings on different aspects of water law and laws,

including comparative approaches and approaches to reform,[11] it may be helpful to have a framework within which to analyse existing laws and future reform proposals.

Law implements policy, and as water (like environment) is of global as well as national concern, the policy contexts relevant to water management are often developed at global level. The next section will consider these global policy agenda(s), and the key players involved in their creation and implementation.

1.2   GLOBAL POLICY AGENDAS

It is arguable that there is no such thing as a global agenda, in water or any other policy area. Nonetheless, over the last 30 years, it is possible to trace the development of a set of policies, in the fields of both water resources and water services, which inform and shape the emerging legal rules that in turn give effect to those policies. These are not always cohesive; especially at a global level, there are competing priorities. In the domain of urban water services, analysed in Chapter 5, there has been a real dichotomy, perhaps even schism, between those who promote a market solution and those preferring a more traditional social policy. Nonetheless, even in water services, although the policy developments may have been schizophrenic, they have been directional; the absence of basic services for so many people has kept water at the policy forefront. In water resources, there has been more agreement, with the introduction of the holistic concept of integrated water resource management (IWRM); yet here, as the practice has developed, there have been questions over the efficacy of the theory. These debates will be explored in later chapters; but first it is worth examining developments in the global arena and identifying in the process some of the actors and organisations that have played key roles.

1.2.1   Networks, agencies and actors

Key stakeholders nationally include national and local governments, agencies of the state, relevant professionals and civil society groups, but also those whose livelihoods depend on water, or who struggle to gain access to basic services. Their engagement will be important to later chapters in this book; but in terms of global policy, there are other players whose roles and interests should be noted.

The United Nations (UN) is the primary global agency, and it has already brought together a set of UN agencies and external partners under the umbrella of 'UN-Water', which is responsible for the World Water Assessment Programme and the World

---

[10] UNEP (2008).

[11] See, e.g., Dellapenna and Gupta (2008), Hodgson (2006), Salman and Bradlow (2006), Bruns et al. (2005); and see also UN-Water (2012).

Water Development Reports,[12] as well as a series of policy papers. The umbrella also covers some of the work on water and sanitation services of the World Health Organization (WHO) and the United Nations Children's Fund (UNICEF).

The World Bank funds many water projects. In the 1990s the Bank, in tune with the prevailing political consensus, stressed the use of market models, private sector participation and competition,[13] but in 2003 it revisited its high level strategy on water,[14] and in the last decade other circumstances have tended to modify such a theoretical stance, especially the move away from investing in long-term concessions in water services.[15] Although, unsurprisingly, much of the Bank's analysis is economic, there is a constant theme of the need for better legal, institutional and regulatory mechanisms, both for service provision and in the management of the resource, particularly water rights and allocation and especially in large infrastructure projects. The development banks have been active in the governance agenda (Chapter 2); water services will be explored in Chapter 5.

Non-governmental organisations play a major role in water. As well as the international environmental groups (e.g., the World Wide Fund for Nature (WWF) or the International Union for the Conservation of Nature), there are a number of new global NGOs specifically concerned with water resources. In the 1990s, after the Dublin Conference (below), two international organisations were established, the World Water Council, an international 'think tank',[16] and the Global Water Partnership (GWP), with a mandate to support water resource management in developing countries. The GWP is a lead institution on IWRM and provides policy advice and guidance through regional partnerships.[17] The World Water Council has been active in organising the World Water Forums; to date there have been six of these. Whilst the specific themes have been different, the general concerns have remained very similar: participation and capacity building; safe clean water for all; institutional, technical and financial innovation. In the Forums, as in the work of the GWP, the need to reform institutions and laws has been a recurring issue.

This section has identified just a few of the organisations and institutions involved on the international stage, but there are many others, governmental, professional and civic; whilst they may contribute to data and to policy, it is also arguable that there are too many players, that their efforts are diffuse and the results sometimes indifferent.[18] Further, whilst in the mid twentieth century the emphasis was on the hydrological sciences and identifying the physical resource base, now it has shifted to a 'softer', governance and management agenda. Both are important, but all the management principles in the world are unlikely to substitute for an understanding of how much water there is in a basin. Although this book looks at legal frameworks, the inter-disciplinary nexus with the water sciences is fundamental if those frameworks are to be properly designed.

1.2.2 Policy developments

Whilst it is feasible to trace modern international policy statements on water back to the Stockholm Declaration in 1972,[19] or the Mar del Plata conference in 1977,[20] this analysis will begin in 1992, when the UN Conference on Environment and Development[21] produced *inter alia* Agenda 21.[22] Agenda 21 devoted a chapter to freshwater resources, whereby state signatories agreed to take action in areas including water resource management, allocation, pollution control and the supply of water services – the four substantive topics of study in this book. In each set of actions, there was recognition of the need for reform of the legislative and regulatory environment.

Agenda 21 was preceded by the Dublin International Conference on Water and the Environment, which had resulted in the 'Dublin Statement'.[23] This set out four principles: that freshwater is a finite and vulnerable resource; that its development and management should be based on a participatory approach; that women play a central part in water management; and that water has an economic value in all its competing uses and should be recognised as an economic good.

These were subsequently reformulated into three principles by the World Bank: the ecological principle – river basin management, environmental protection, and managing land and water together; the institutional principle – subsidiarity and the inclusion of all stakeholders; and the instrument principle – a scarce resource requires incentives and economic instruments to manage effectively.[24] These are sometimes described as the IWRM principle, the 'decentralisation' (or participation) principle, and the 'privatisation' or economic principle. As regards the last, it is important to note the recognition in the Dublin sub-text that firstly there is a basic right of access to

---

[12] UN-Water (2003, 2006, 2009, 2012a). In future, these will be annual, and targeted.

[13] See, for a trenchant critique of the World Bank approach to water, Finger and Allouche (2002), especially Chapter 3.

[14] World Bank (2004).   [15] Marin (2009).

[16] 'World Water Council' see generally http://www.worldwatercouncil.org/index.php?id=1.

[17] 'GWP' see generally http://www.gwp.org/.

[18] See, for a critical analysis, Varady and Iles-Shi, 'Global Water Initiatives: What do the Experts Think?' in Biswas and Tortejada (2010).

[19] UN (1972). The Stockholm Conference agreed that states had a right to exploit their own environment, but also a responsibility to other states; still a founding principle of modern environmental law.

[20] UN (1977).   [21] UN (1992).

[22] UN (1992a) (Agenda 21). Chapter 18 specifically addresses freshwater resources.

[23] Dublin Statement (1992).   [24] World Bank (2004).

water. Otherwise the 'special nature' of water risks disappearing in a purely economic analysis of service provision and cost recovery, at the expense not just of the basic human needs of those who cannot pay, but also of ecological needs, and of what might best be described as the spiritual aspects of water. This special nature is reflected in the European Community's Water Framework Directive: 'Water is not a commercial product like any other, but, rather, a heritage which must be protected, defended and treated as such.'[25]

The debate around the Dublin Principles has been dominated by principle four, and the promotion of the market-oriented approach; this has fostered the schism in the debate around water services. It has also significantly affected approaches to the management of the resource, including IWRM and reform of water rights, such as the ideologically driven reforms of water markets in Chile,[26] or developments in India.[27] The need for law reform is still apparent.

The Johannesburg Summit on Sustainable Development, 10 years after Rio, took forward the global sustainable development agenda with the emphasis on delivery rather than new policies.[28] There was a specific requirement for all signatories to produce IWRM and water efficiency plans at all levels by 2005. There was also provision for better water pollution control, recognising that this benefits public and ecosystem health. Efficient use and better mechanisms for access and allocation were called for, and water and sanitation issues were still a priority. All of these policy areas are relevant to the analysis in later chapters of this book. All the policy documents surveyed above make mention of stable and transparent regulation as one tool for better management.

1.2.3 The Millennium Development Goals and Sustainable Development Goals

At the start of the twenty-first century, the broad policy objectives received new focus with the production of the Millennium Development Goals (MDGs).[29] Goals in relation to water included halving the proportion of people without safe drinking water, or access to basic sanitation,[30] by 2015. Water is recognised as a cross-cutting issue, relevant to all the MDGs. In the most recent reports, the drinking water target is being achieved, but not that for sanitation, with the biggest deficits in sub-Saharan Africa and southern and eastern Asia.[31] The MDGs will not all be realised by 2015, and the international

community is taking the agenda forward following the 'Rio +20' Summit in 2012.

This took place in a very different political and economic environment. Following the global financial crisis, and the failure to meet many of the MDGs, it is perhaps not surprising that the international community has not shown the common cause and purpose that seemed evident in the outputs of Rio in 1992, or even Johannesburg in 2002. The 'outcomes' document from 2012 is very different, and relatively limited.[32] It reaffirms many existing high level commitments, including the water and sanitation MDGs, and sustainable development and poverty eradication. It emphasises the importance of good governance, and of human rights, including the human rights to water and sanitation. In the few paragraphs on water, there is commitment to the progressive realisation of these rights, as well as the role of ecosystems, the need to manage water pollution and treat wastewater, the management of flood and drought and the use of non-conventional water sources.[33]

The international community is now considering what should be done to take forward the work of the MDGs after 2015, including the creation of sustainable development goals. In water, there have been three thematic sub-groups: water, sanitation and hygiene; water resources management; and wastewater and water quality. At the time of writing, this process is still under way, but a report has been produced.[34] It stresses the need to move away from narrow goals and silos, build collaboration, and recognise that water will continue to cut across all development and poverty-alleviation activities. It suggests ambitious goals and targets, including universal access to basic services, and further that a rights-based approach to water needs to move beyond a narrow perception of water and sanitation and recognise policy interlinkages, especially with food, as well as the inter-generational principle of sustainable development. The relationship between water and other critical sectors – the water/food/energy nexus,[35] and the multiple impacts of climate change – is identified. So too is the need to address water for nature, to ensure the continuation of the services that ecosystems provide. On wastewater and water quality, there is recognition that a combination of urbanisation and population growth means we are all downstream users now. The report urges the collection and treatment of wastewater; as with solid waste, there is a critical need to manage this as a valuable resource base, and to overcome some of the taboos and negative perceptions which, as with sanitation, move this issue too far down the policy agenda. As might be expected, there is recognition of a growing debate around water security (itself a term with many meanings);[36] the

---

[25] Directive 2000/60/EC (WFD), Preamble.   [26] Bauer (2004).
[27] See, e.g., Olleta 'The Role of the World Bank in Water Law Reforms' in Cullet et al. (2010).
[28] UN (2002).   [29] UN General Assembly (2000).
[30] The sanitation goal was introduced at the Johannesburg Summit, UN (2002) para. 25; and see also Chapter 5.
[31] WHO/UNICEF (2012).

[32] UN (2012).   [33] UN (2012) paras. 119–124.
[34] UN-Water/UNDESA/UNICEF (2013).
[35] See, e.g., Bonn Nexus (2011), UN-Water (2012).
[36] Wouters (2010), Magsig (2013).

need for governments to work with many stakeholders; the need for capacity development; and, of course, the need for finance.

Whilst it would be possible to write more extensively on these policy formulations, the ends, if not necessarily the means, have a degree of consistency. Provision of drinking water and sanitation, access to water for other uses especially agriculture and food, the links to economic activity, personal and public health, and societal wellbeing are all prominent, as is the need to protect both surface and groundwater from over-extraction and pollution. The problem is not a lack of freshwater, but the failure to manage that water effectively to provide for the needs of the global community, by the application of adequate funding, backed by political will.

1.3 SCOPE AND APPROACH

This book provides a legal analysis, taking a comparative approach with reference to primary materials, principally national and supranational legislation, and policy documentation both national and international. There is no intention to develop a single or best model; it does not aspire to provide a normative framework. However, each chapter has some normative content, identifying the policy context(s) for the development of the law within each of the core areas. These contexts, at least to an extent, prescribe norms of conduct, and/or the values that underpin them, by setting policy goals that states and others should achieve, *inter alia* through regulation. The goals of efficiency, equity and environmental (or ecological) sustainability – the 'three E's' of water management – are predominant in this regard, and may be expressed as principles or purposes within legislation. To this extent the book supports the approach of the Realist school[37] in recognising the interdependence of the law and of institutional arrangements within the broader social and economic milieu, but it does not purport to provide a sociological or economic analysis *per se*, any more than it is an empirical study. Similarly, it will make reference to the role of other disciplines in policy formulation and legal development, without claiming to be an interdisciplinary study.

The analysis of the policy context has normative elements but the analysis of the law is predominantly positivist. It examines the law as it exists, but to an extent it also considers both the practice of its implementation and the intentions of the policy-makers and the legislators. This last in particular will connect the subsequent legislation to the policy context, and some conclusions will be drawn as to the success or otherwise of achieving the policy goals through the various options considered, but there

is no systematic attempt to make value judgments about the extent to which the policy goals have been reached, as the objective is to examine various options, all of which may be seeking to achieve the same or similar results.

Whilst the analysis of the law and practice is essentially positivist, albeit contextualised, it is also reflexive. Analysis of the law in force is made in the context of the policy drivers, and the conclusions consider the structural elements of a reformed law that will be essential to meet the policy goals, as well as (in part) the ability of a particular option or legal model to achieve these goals. The underpinning legal philosophy is the concept of pragmatic cosmopolitanism.[38] This recognises the increased globalisation of law, and the impact of global agendas on national regimes – the concept is closely linked to transnationalism, or transnational law. It is also fundamentally pragmatic, as it seeks to analyse what the law is and what it can be; it is aspirational, certainly, insofar as there are normative elements, but essentially it is intended to be realistic, and grounded in practice and achievability.

1.3.1 Scope: what is 'water law'?

Before proceeding to the substance, it is perhaps useful to consider briefly the scope of this book, in two aspects: firstly, what is included in 'water law'; and, secondly, some discussion of the choice of jurisdictions, along with some supporting information about those countries, their legal systems and constitutional arrangements, and their water use.

This book looks at the components of a reformed 'water law', but what is 'water law'? The core elements identified are water resource management; water rights and allocation; water quality and pollution control; and water services, here used to mean the supply of drinking water, wastewater and sanitation services (often described as urban water services). Each merits a separate chapter. The first is described herein as strategic, and the others as functional, or operational.

These choices may be obvious, but there are other operational control regimes pertaining directly to water. In addition there are other strategic regimes, such as land use planning, which affect management of the water environment and support its reform, as well as many sectors whose activities affect the resource and may have their own separate legal provisions.

Water resource management, along with rules on abstraction and pollution, forms a coherent whole which may be reformed within a single legislative framework. Water services are not usually an integral part of such a unified reform package, and it is not necessarily, or indeed usually, desirable to reform water services within the same legislative framework or at the same time. Further, it is arguable that urban water services are a

---

[37] The US school; especially, the work of Oliver Wendell Holmes, Karl Llewellyn and Jerome Frank. For an introduction, see, e.g., Freeman (2001), Chapter 9.

[38] See, e.g., De Waal (2005), Samuel (2003).

sectoral use of water, and certainly in terms of the proportion of global water use it is far less significant than irrigation water (see Table 1.2 below). Nonetheless, there are arguments for addressing water services in this work.

Firstly, the provision of drinking water and sanitation is an area of acute unmet need, as evidenced by the global policy agendas set out above, and this imperative has also driven forward the broader agenda for reform of water resource management. Secondly, the management of irrigation water takes place squarely within the broader water resource management framework for abstractions and water quality control, albeit with a wealth of detailed specialist provision, but regulation of water services brings a different dimension. Thirdly, in developed northern countries, such as Scotland and England, a significant proportion of water used is delivered via the water services providers.

The figures here provide conceptual models for national water law. Figure 1.1 shows what is herein described as the water law meta-regime, with the core operational elements of allocation, pollution and water services, subsidiary to the strategic framework of IWRM. The other operational regimes shown here, such as flood and drought, or coastal and marine waters, still pertain directly to water, but also raise other issues. These will ideally be integrated through a broad IWRM framework; this book will touch on them, but will not analyse them in detail. Figure 1.2 shows other strategic legal regimes that support water management, and also key sectoral uses. Many of the strategic regimes, such as land use planning or environmental protection, would

exist as another meta-regime similar to water law; the environmental law meta-regime is also considered in Chapter 4.

Following this introduction, Chapter 2 will address integrated water resource management, and also links between water law and other regimes. Just as resource management sets the framework for the operational aspects of water law, so Chapter 2 will set the framework for the rest of the book. It will include discussion of governance and stakeholder participation, which are contextual throughout, and consider briefly other strategic regimes, and some other aspects of water law.

Chapter 3 will look at water rights; at abstraction and allocation. This will include some discussion of pre-existing regimes

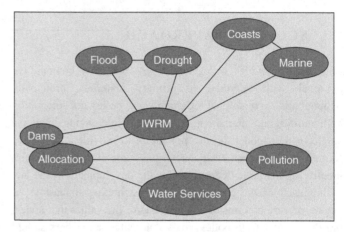

Figure 1.1 Water law meta-regime.

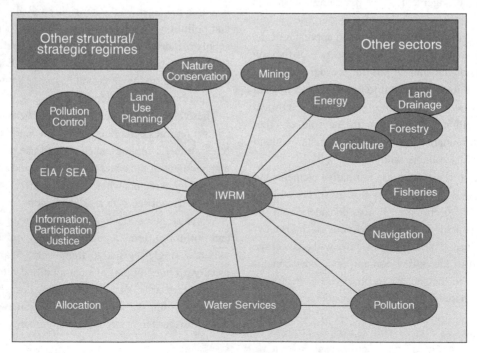

Figure 1.2 Other related legal regimes.

for water rights and water use, particularly riparianism, and the issues surrounding reform of such rights, which may have the characteristics of property rights. It will then address the new provisions for allocation of rights in water, including the status of existing users and whether any exemptions are made from the licensing requirement, e.g., for subsistence use or small abstractions. Licensing regimes generally will be considered in this chapter. It will look at bulk supply and water pricing, and end with discussion of water rights trading, particularly in Queensland.

Chapter 4 will look at water pollution and water quality, in the context of environmental protection meta-regimes in each of the countries involved, and the emerging paradigm of an ecosystems approach. It will address the use of standards or guidelines, and the relative merits of departmental or independent regulators. It will also consider the developing mechanisms for assessing ecological quality, again leading on from the work in Chapter 2. Whilst environmental protection from point sources may be well established, the management of diffuse pollution and ecological degradation are continuing problems for the twenty-first century.

Chapter 5 will consider urban water services. It will look at the debate over the 'human right to water' and consider the relevance of the human rights agenda to meeting basic needs. It will look at the models for water services provision – public sector, private sector and hybrid models – and consider whether and how the components of the service might be disaggregated. Without developing into an economic analysis, it will look at legal structures underpinning regulation of water services, and will assess the functions and duties of providers, finishing with consideration of water conservation and demand management, including wastewater reuse.

Chapter 6 will draw general conclusions as to a framework for reforming water laws.

1.3.2 Scope: the choice of jurisdictions

The choice of jurisdictions is of importance to any comparative study. All of the jurisdictions studied here have either recently undertaken, or are in the process of, major reforms to the legal and management frameworks for water resources, and the specific drivers for these reforms will be an integral part of the analysis.

Whilst Scotland and England are northern countries, with a preponderance of urban domestic and industrial water use and very little irrigation, in both Queensland and South Africa the proportion of water used for irrigation is closer to the global norm. The United Kingdom (UK) jurisdictions are within the European Union (EU), which is a driver for change, but which can also be analysed in its own right as an exemplar of certain approaches to water management,

especially in water resources management and water quality. The Scottish case is interesting here as Scotland has been very proactive in implementing EU water law, going beyond the requirements of EU directives in the national reform programme. Both England and Scotland have very particular models for water services, including a fully divested industry in England and a highly regulated public provider in Scotland.

Australia as a whole provides many options for water law and management, including a developed water trading regime within a federal system, and since the 1990s there has been a series of Commonwealth policy initiatives relating to the environment and to water, which will then be transposed into state legislation. In many ways it is these Commonwealth initiatives that make Australia an exciting and relevant comparator for this book. However, as water is a state function, it is also necessary to select a particular state. When the original research for this work was done, in 2002–2006, Queensland was selected as it was implementing Commonwealth reforms somewhat later than other states, and benefiting from their experience. In the intervening period, some aspects of the state law have been extensively reformed, and this process is ongoing, which presents challenges, but also makes for some interesting analysis as to the purposes of the various reforms. To an extent, this is also true in England. The law is rarely stationary and at the time of writing all of the jurisdictions are making some new changes, which will be considered as appropriate. It should be noted here that, generally, all references to legislation are to the current amended versions of the principal rules, unless there is a reason to specify the amending rule. Similarly, as departments may change their names and functions, these will generally be referred to in the text by their current name, unless the context requires otherwise; but documents will be cited using the name of the department as it was when the document was written.

South Africa brings lessons for both developed and developing countries, and has been written about and commented on extensively. The post-apartheid reforms led to a complete review of all aspects of water law in a situation with a real political will for change, and South Africa is also a major regional influence. As a country with an arid or semi-arid climate, huge variability in wealth and in access to both resources and services, and a predominantly rural subsistence economy, it provides many contrasts to all the other jurisdictions. In water services in particular, it is important to consider a jurisdiction where at least some of the population share in the current crisis in services provision in the developing world.

Between these jurisdictions there is sufficient variety to provide meaningful comparisons, whilst in each of them, at least some aspects of their water laws are capable of being considered a worthy example for others to consider.

1.3.3 Relevant constitutional arrangements

The United Kingdom of Great Britain and Northern Ireland consists of four countries – England, Wales, Scotland and Northern Ireland. The provisions of the Scottish Act of Union[39] were such that Scotland has retained her own legal system, and separate system of private law; and in areas affected by the historic private law, including property law and therefore water rights, Scots law has developed differently from that in England. The UK Parliament has sovereignty but unlike the other jurisdictions (indeed unlike almost any other country in the world) has no written constitution. Recent devolution has given Scotland a new Parliament,[40] which holds devolved powers regarding the environment, private property rights, water, and the implementation of relevant EU law. At the time of writing, there is to be a referendum on Scottish independence, in autumn 2014.

The UK is a member of the EU,[41] a regional organisation with a highly developed supranational legal system. EU law must be applied by Member States; it has supremacy over national law, and the EU has legislated extensively in the field of the environment and water. EU water law will be analysed throughout this book.

The Commonwealth of Australia comprises six states and two major territories, and was established by the Constitution of Australia Act 1900.[42] This sets out the powers and functions of the Commonwealth (also known as the Federal Government or the Government of Australia); any functions not specified are state functions. Naturally, in 1900 no mention was made of the environment. The Commonwealth has competence in external affairs and, as the state entity for international law purposes, is the signatory to international conventions, declarations etc.; in that case there may be legislation implementing those agreements at Commonwealth level. It acts in the field of the environment where there are issues affecting the whole of Australia, and often in conjunction with New Zealand; there are a number of Ministerial-level bodies establishing policy across both countries. There is a Council of Australian Governments (COAG) which initiates policy reforms in areas that affect all states, including aspects of water reform.[43]

The Republic of South Africa rose from its apartheid past with the first free and fully franchised elections in April 1994. Subsequently, a draft constitution was consulted upon, reviewed, approved by the constitutional court and came into effect in 1997.[44] The Constitution has many model features including a founding principle of cooperative government,[45] and specific rights to a clean environment[46] and to water.[47] There are nine Provinces, and the Parliament consists of both the National Assembly and the National Council of Provinces.[48] In addition, there are metropolitan and district municipalities. The principle of cooperative government leads to some overlap for responsibilities in the field of the environment, but this is less problematic for water, where resource management is a national function, whilst water services are provided by local government.

1.3.4 Country data and analysis

Table 1.1 gives some general information on land area, population and available water resources for the jurisdictions under review, to provide a context for the study of their water resource management provision.[49]

The information given demonstrates the disparities. South Africa's land area is comparable to that of Queensland, and both have arid areas and large expanses of land with low population; Queensland's population density is extremely low; England's is significantly higher than any of the other comparators. All jurisdictions have variable rainfall but Queensland's is the most extreme; the northern wet tropics have the highest rainfall in Australia, higher than the west coast mountains of Scotland. High rates of evapotranspiration mean that very little of South Africa's runoff reaches the sea. Regarding available water resources, countries with less than 1700 m^3 per capita are

---

[39] Treaty of Union 1706; Act of Union 1707 c.7.

[40] Scotland Act 1998 c.46; this has significantly increased the scope for law reform, after many years of limited Parliamentary time at Westminster for Scottish matters. The Scotland Act 2012 c.11 extends the devolution settlement, pending the referendum result.

[41] Since the European Communities Act 1972 c.68.

[42] Constitution of Australia Constitution Act 1900 63 & 64 Vict. c.12, as amended.

[43] COAG was initiated in 1992 and comprises the Prime Minister, State Premiers, Territory Chief Ministers and the President of the Australian Local Government Association. In water reform it has been particularly concerned with competition policy, water rights and water trading, and will be of relevance to many aspects of this book.

[44] Constitution of South Africa Act No.108 of 1996.

[45] Constitution of South Africa ss.40–41.

[46] Constitution of South Africa s.24.

[47] Constitution of South Africa s.27.

[48] Constitution of South Africa s.42.

[49] The information in Table 1.1 is taken from the following sources: 'Australian Bureau of Statistics' http://www.abs.gov.au/; 'Australian Government Bureau of Meteorology' http://www.bom.gov.au/climate/current/annual/qld/summary.shtml; 'Australian Government Geosciences Australia' http://www.ga.gov.au/education/geoscience-basics/dimensions/area-of-australia-states-and-territories.html; Government of South Africa (2013); 'Government of South Africa: About South Africa' http://www.gov.za/aboutsa/geography.htm; 'Population Estimates Scotland' http://www.gro-scotland.gov.uk/files2/stats/population-estimates/mid2012/j29078400.htm; 'Queensland Government Statistician's Office' http://www.oesr.qld.gov.au/products/briefs/pop-growth-qld/qld-pop-counter.php; 'Scotland Info' http://www.scotlandinfo.eu/weather-climate.html; 'UK Government Office of National Statistics' http://www.ons.gov.uk/ons/taxonomy/index.html?nscl=Population; 'UK Meteorological Office Climate and Rainfall' http://www.metoffice.gov.uk/climate/uk/actualmonthly/; 'World Bank Renewable Internal Water Resources' http://data.worldbank.org/indicator/ER.H2O.INTR.PC.

Table 1.1 *Country data*

	South Africa	UK	England	Scotland	Australia	Queensland
Land area (km^2)	1,219,090	241,930	130,422	78,772	7,659,861	1,723,936
Population (millions)	51.78	63.23	53	5.29	23.4	4.72
Population density (/km^2)	42	261	406	67	3	2.7
Long-term average rainfall (mm/annum)	450	1160	840	1560	486	623
Long-term average rainfall variability (mm/annum)	$<200->600$	$<600->3000$	$<600->1200$	$<800->3000$	$<200->4000$	$<200->4000$ m
Resources per capita (m^3)	886	2311			22,039	

considered to be water stressed, and those with less than 1000 m^3 per capita are water scarce.[50] The Australian figure is distorted by the northern tropics, and the low population density.

In the UK, Scotland includes the wet north and west highlands, and the relatively flatter and drier east coast. In England, the southeast is considerably drier than other regions and also has a very high population density. Neither jurisdiction has the same extremes of climate as South Africa or Queensland. The Gulf Stream, bringing warm water to the western coasts of the British Isles, keeps temperatures significantly warmer than would be expected at such northerly latitudes. The British Isles have a variety of aquatic ecotypes, but no great rivers as are found in the Americas or Africa. Water resources are not stressed in the UK as a whole, but are in the southeast of England.

In Australia, Queensland runs down the east coast from the wet tropics in Cairns and further north, to Brisbane and the Gold Coast above New South Wales, with the Murray–Darling river system as its southern boundary. The sparsely populated western hinterland towards South Australia is arid desert, and the bulk of the expanding population live in the greater Brisbane area in South East Queensland. The Murray–Darling is the only significant river system in Australia and its management will be considered in Chapter 2.

In South Africa, there is great variety of climate from the arid desert in the northwest towards Namibia, to tropical forest on the east towards Mozambique. The majority of the population live on the coast and in the east, whereas the northwest is sparsely inhabited. Many of South Africa's rivers have intermittent flow, and only the Orange and the Limpopo maintain permanent channels to the sea. South Africa has land borders with five states and also encloses the Kingdom of Lesotho, and has international agreements with all of these regarding water.

Table 1.2 gives some comparative data regarding water use.[51] Figures on sectoral water use are difficult to obtain and often inconsistent; for example, results will vary on whether industrial use includes water for cooling and for hydro; whether agricultural water includes water for fisheries, or water delivered through the mains as well as water directly abstracted; and whether urban domestic water (which may also be called municipal water, or water delivered as public supply) includes mains water for industry (and indeed agriculture).[52] A range may be more reflective of the debate, which often concerns measurement, monitoring and assessment as much as analysis of the results. This points again to one of the underlying themes of this book, which is the often difficult relationship between policy-into-law and the scientific evidence base, which ideally should underpin that policy and hence the emerging law.

In Queensland and South Africa, withdrawals for agriculture (primarily irrigation) are comparable to global averages. In the UK, the low proportion for agriculture reflects both the proportionately high industrial use and the preponderance of rain-fed farming. There is some irrigation, especially in the south of England (as much as 16% of withdrawals in East Anglia) and to a very limited extent in the northeast of Scotland; flooding and land drainage are also major localised issues. Regional figures are so variable, and so difficult to compare, that it was decided not to attempt to give values for England and Scotland. In Queensland, rural domestic use is usually supplied via irrigation networks and therefore may be included in agricultural use and not as municipal supply. Per capita use reflects differing global norms – Australia generally has very high levels of domestic consumption, similar to the USA, though Queensland is lower than other states; the UK is still a middle-ranking consumer. In South Africa, the variation is more informative than the average, with the rural poor subsisting on marginal consumption, and the richest citizens consuming as much as anyone in the developed world; the upper bound cited is probably an underestimate.

---

[50] UNDP (2006) p.135.

[51] Figures in Table 1.2 are taken from the following sources: EA (2011); DERM (2012); Earle *et al*. (2005); 'South Africa Water Resources Council' available at http://www.wrc.org.za/Pages/Resources_Regionalstats.aspx; UN-Water (2009); WaterWise (2007); 'World Bank Annual Freshwater Withdrawals' available at http://data.worldbank.org/indicator/ER.H2O.FWAG.ZS/countries.

[52] For a discussion of the difficulties, see Krinner *et al*. (1999) Chapter 3.

This chapter has set out some of the key issues affecting management of water, and the policy contexts that drive law reform. It has attempted to show the relevance of water law, its relationship to other legal regimes, and the need for frameworks to guide reform. It has set out some basic information about the jurisdictions under review, and the structure of the chapters to follow. Chapter 2 will now proceed to assess the legal frameworks for the strategic goal of integrated management of water resources.

Table 1.2 *Sectoral water use*

	South Africa	UK	Australia	Queensland	Global
Agriculture	57–67%	3–16% (regional variation)	74%	62%	70%
Domestic	22–31%	40–57%	16%	12%	10%
Industry	6–11%	33–45%	11%	26%	20%
Groundwater use	13%	7–33% (regional variation)	31%	35%	20%
Per capita use	<50 – >250 LPD	145 LPD		220 LPD	

2 Integrated water resource management and river basin planning

2.1 INTRODUCTION

The proper management of resources is part of the sustainable development agenda: in Chapter 1, we saw that since the Rio Summit global policy for water has focused on the concept of IWRM. This has been defined by the GWP as 'a process that promotes the coordinated development and management of water, land and related resources, in order to maximise the resultant economic and social welfare in an equitable manner, without compromising the sustainability of vital ecosystems'.[1] Thus IWRM should encompass land as well as water management, and address social, economic and environmental concerns; the need to consider the whole resource, especially surface and groundwater, to address both quantity and quality, and to give a voice to water users, are implicit. Since Rio, the question of stakeholder participation and the mechanisms for taking forward decision-making processes – not just in IWRM, not just in water – have often been framed in terms of a 'governance' debate, and this chapter will also consider that concept, in the context of structures for participation and engagement.

IWRM has been subject to criticism, as being too complex, too expensive, or too difficult to assess in terms of its concrete results.[2] Even from a purely legal perspective, a state with limited resources and severe water allocation problems, for example, might be better placed to concentrate on reform of water rights and abstractions, rather than focusing on large-scale governance and participation initiatives that might come from notions of IWRM.[3] In addition, the economic focus of the Dublin principles, and the approach of the World Bank in funding projects, is considered to have skewed approaches to water management (both resources and services) in a way unhelpful in developing countries.[4] Yet Lenton and Muller suggest that the criticisms of IWRM are not well founded, and that what is needed is a return to the essence of Agenda 21.[5] Muller has gone

further and noted that Agenda 21 addressed integrated water resources development and management, and there should be more attention to development.[6]

It is also arguable that concentration on 'soft' management concepts deflects attention from physical and hydrological imperatives; there is no point in complex systems for governance and participation if the availability of water in the basin is unknown. The first task in water management must be to identify and assess the water available, along with its current and likely future uses, and to establish adequate monitoring of quantity and quality; this chapter, and this book, start from this premise.

The other essential elements of IWRM are an integrated approach to the water cycle, the engagement of water users at appropriate scales, and integration with non-water policy areas, especially land use. Chapter 1 has already outlined the water law meta-regime, and identified IWRM as the strategic level, coordinating with (or integrating) the operational areas – abstraction, pollution, services – and providing a route to link to other related sectors and strategic regimes. If a state does choose to reform, say, its water rights regime, it can do so without first setting up an administratively complex and burdensome system of water resource planning; but some of the elements that would be established in a legal framework for IWRM are likely to have to be located instead in the new structure for that operational area.

Another general question that arises is the relationship between IWRM and 'river basin planning'. In some ways they can be seen as synonymous. The hydrological interdependency that requires the co-management of surface waters and groundwater also recognises the hydrological boundaries of the basin, catchment or watershed, and the difficulties that can arise when these physical units are divided across administrative or indeed national boundaries. Agenda 21 suggests that IWRM 'should be carried out at the level of the catchment basin or sub-basin'.[7] However, both scale and country-specific needs are highly relevant to the choice of management unit. The great transboundary basins covering territory in many different states, such as the Nile, will also need organisational structures at much smaller scales; whilst very small catchments are unlikely to have

---

[1] Rogers and Hall (2000).
[2] See, e.g., Biswas (2004), Biswas (2008), Watson *et al.* (2007).
[3] See, e.g., Hodgson (2006).
[4] See, e.g., Zodrow, 'The Role of the World Bank in Water Law Reforms' in Cullet *et al.* (2010).
[5] Lenton and Muller (2009), Chapter 14, for a review and response to the critics; see also UN (1992a) para.18.5.

[6] Muller (2010).   [7] UN (1992a) para.18.9.

the resources for major planning initiatives. In South Africa, the overarching strategy for water has been produced at national level.[8] So again there is no prescription as to exactly how states organise their water management; nonetheless, most of the public agencies and NGOs involved in water consider the river basin to be an important management unit.

Given the widespread acceptance of the concept in the last 20 years, it is unsurprising that there has been a significant amount of policy guidance produced to assist states, water managers and other stakeholders. The GWP has *inter alia* produced a 'tool box' – an open source database of guidance, examples and case studies;[9] they have also recently co-published a handbook on IWRM.[10] UNESCO, through its International Hydrological Programme, has also produced guidelines on practical implementation of IWRM,[11] which was a side publication to the Third World Water Development Report, and these are designed to be complementary to the GWP work, not overlapping. The GWP publication was co-produced with the International Network of Basin Organisations, a global NGO network, whilst UNESCO itself runs a cross-cutting programme for river basin management, 'Hydrology, Environment, Life and Policy', which also takes a broad approach to management and seeks to engage stakeholders and break down barriers, both horizontal and vertical.[12] There is no absence of information, advice and supporting networks. There has also been a recent analysis by UN-Water on the extent to which countries have succeeded in implementing the commitment in the Johannesburg Declaration to introduce IWRM,[13] suggesting that most countries have undertaken some steps towards this policy goal, including legislative and policy change (around 80%) and IWRM planning processes (around 65%), although implementation is falling behind for the least developed countries.

If IWRM is clearly part of the global policy construct, despite the criticism, some commentators now take the view that it is also emerging as an underpinning principle of modern water law.[14] In keeping with the broadly positivist and pragmatist approach of this book, there is no presumption here that IWRM is mandatory, but nonetheless each of the jurisdictions herein reviewed does provide for IWRM, and the analysis supports the identification of IWRM as established policy and as an emerging legal principle. This chapter will analyse comparatively the structures and institutions, and specific legal provision for IWRM, in each jurisdiction, along with the provision made in

law for participation, and links to the other structural and strategic regulatory frameworks that will be relevant throughout this book.

Just as IWRM is not a goal in itself, neither is the production of a plan, policy or strategy. Nonetheless, it is very likely that such documentation will be produced, as a mechanism for presenting data and information, representing the outcomes of decision-making, taking forward the various operational activities, and linking water to other resource management regimes, especially land use. In each of the jurisdictions there is some form of planning process and the law can, and should, make appropriate provision for that.

2.2 TERMINOLOGY, DEFINITIONS AND SCOPE

In some ways, the law is all about definitions; these set boundaries and scope, and hence give clarity. Until we define the relevant terms, we cannot be certain what activities will be controlled, or in what parts of the water environment, in any one jurisdiction. The terms 'river basin', 'catchment' and 'watershed' are widely used in the literature, are often used synonymously, and may also be used in legal instruments. Thus the EU Water Framework Directive (WFD)[15] defines 'river basin' as 'the area of land from which all surface run-off flows through a sequence of streams, rivers and, possibly, lakes into the sea at a single river mouth, estuary or delta'. 'River basin district' (the administrative unit for water management under the WFD) is further defined as 'the area of land and sea, made up of one or more neighbouring river basins together with their associated groundwaters and coastal waters...' Transposing legislation in England and Scotland uses the same definitions for WFD terms; in addition, in Scotland the Water Environment and Water Services Act (WEWS)[16] defines the 'water environment' as also including wetlands, which brings these into the control regime for abstractions, discharges, and other activities.

In South Africa, the National Water Act (NWA)[17] defines 'catchment' as 'in relation to a watercourse or watercourses or part of a watercourse, means the area from which any rainfall will drain into the watercourse or watercourses or part of a watercourse, through surface flow to a common point or common points'. The term 'water resource' is defined to include groundwater ('aquifers') as well as watercourses and estuaries (all of which are further defined), but not coastal waters. 'Watercourse' specifically includes wetlands, and also temporary flows, and any 'collections of water' so declared. In Queensland, the

---

[8] DWA (2013).   [9] 'GWP ToolBox' see http://www.gwptoolbox.org/.
[10] GWP/INBO (2009).   [11] UNESCO (2009).
[12] 'Hydrology, Environment, Life and Policy' see http://www.unesco.org/new/en/natural-sciences/environment/water/ihp/ihp-programmes/help/.
[13] UN-Water (2012).
[14] See, e.g., Salman and Bradlow (2006) Section 3.2. For an American perspective, see Guruswamy and Tarlock, 'Sustainability and the Future of Western Water Law' in Kenney (2005).

[15] Directive 2000/60/EC (WFD). All definitions are in Art.2.
[16] Water Environment and Water Services (Scotland) Act 2003 (WEWS) asp.3 s.3.
[17] National Water Act 1998 No.36 of 1998 (NWA). Definitions are in s.1.

Water Act 2000 (QWA)[18] defines 'water' and also 'watercourse', 'underground water' and 'overland flow' (diffuse surface water), but the QWA does not define 'catchment' or 'river basin'. In much of the wider literature, the terms catchment and river basin are used interchangeably, and in this chapter will be considered broadly synonymous, but with recognition that for formal legislative and management purposes, South Africa and Queensland are likely to use the word 'catchment' for management at a larger scale, whereas in Scotland and England, this will normally refer to much smaller sub-basin units or small river systems. In the analysis of any legal structure, the precise definitions given to technical terms will always be important; the definitions of the 'water environment', or the 'water resource', will be critical in determining the scope of control of both quantity and quality of water under the law.

Two further preliminary issues are highly relevant to the scope of a new water law in general, and specifically to an IWRM process. One is the authority of the state to legislate for water, and the other is the set of principles that underpin that legislation. The first raises important questions around the ownership and control of the resource. In general, states have sovereignty over their natural resources and a right to exploit the same, subject to the interests of other states;[19] this would include inland and territorial waters. Where any part of those resources is capable of being held in private ownership, as is usually the case for land, there is always a balance between the public and private interest at national level, and this may constrain the state's right to regulate. This question of ownership and control is directly relevant to Chapter 3, on water rights and allocation, and will be considered more fully there, but if a new water resources law is being created and deals with both IWRM and allocation, and if the control of natural resources is not already addressed elsewhere, perhaps in a constitution, then it should be part of that water law. Similarly, it is likely that there will be certain overarching principles that will underpin the application of the detailed rules of water management, whether strategic or operational, and again these may be appropriately stated in a water resources law, if they are not already in place in some other instrument.

2.3 GOVERNANCE, PARTICIPATION AND JUSTICE IN WATER MANAGEMENT

Governance is a broad political concept, and can improve weak or ineffective government or, alternatively, give a voice to the disenfranchised in an unrepresentative system.[20] The GWP has examined water governance as a set of political, social and economic structures, relevant to both water resources and water services, at different levels.[21] The development banks have identified relevant elements or components of good governance, including accountability, transparency, participation and predictability.[22] The UNDP added the rule of law, responsiveness, consensus, effectiveness and efficiency, equity and strategic vision.[23] These analyses are not focused on water or on law, but one major EU study did consider water governance and, within that, ways in which the law can frame and promote the concept.[24] The conclusions were that three key 'process' elements are critical, namely, transparency (access to information), participation and accountability (access to justice). These are features that a good legal structure can enable; they are highly relevant to IWRM and also to the governance of water services (Chapter 5).

These might in the past have been described as a broad participation agenda, especially relevant to environmental management (and environmental law) and to land use planning, for example through environmental impact assessment (EIA). Since the 1960s, there has been an emerging recognition that citizen participation should be a vital component of decision-making in these spheres.[25] This was given more impetus following the Rio Conference in 1992; the overarching principle of sustainable development has always included citizen participation in its social dimension.[26] Equally, sustainable development has underpinned IWRM as it has the broad field of environmental management, and there are many cross-overs between sustainable development and governance. Both are a 'catch-all', meaning different things to different people; but both are important, encapsulating ideas of benefit that have taken forward the debate. Sustainable development broadens environmental regulation to encompass resource management; governance broadens our perception of the appropriate sets of actors and roles.

Linked to both, and to that broad participation theme, is the notion of environmental justice. The three elements of participation can be depicted in various ways: participation in a broad sense can encompass both access to information and access to justice, as well as participation *per se*, for example opportunities to respond to or engage in a decision-making process. It can also be seen as a linear process, with transparency and information flow at the start, leading through opportunities to engage, and ending with the ability to hold the decision-maker to account. These ideas are not exclusive to law or to lawyers, but undoubtedly legal frameworks can and should facilitate them. The

---

[18] Water (Qld) Act 2000 No.34 (QWA). Definitions are in Sch.4.
[19] UN (1972).
[20] See for a general discussion, Stoker (1998); for water, see, e.g., Biswas and Tortajada (2010).

[21] Rogers and Hall (2013).
[22] World Bank (1992); Asian Development Bank (1995).
[23] UNDP (1997).   [24] Rieu-Clarke and Allan (2008).
[25] Arnstein (1969).   [26] UN (1992) Principle 10.

Aarhus Convention is a prime example of an international legal instrument that does so provide.[27]

Public participation generally is one of the strategic regulatory regimes identified in Chapter 1 as supporting the reform of the water law meta-regime, and these are not explored as such in this book. However, stakeholder participation is also a fundamental component of IWRM[28] and is discussed accordingly for each jurisdiction. This is an area where a robust legal framework can be of great benefit, identifying stakeholder groups, providing for their engagement with the planning process, and requiring the decision-makers to take account of different inputs and show how they were considered.

2.4 THE EUROPEAN UNION AND THE WATER FRAMEWORK DIRECTIVE

Of all the jurisdictions and institutions studied in this book, the EU has been very forward-looking in terms of IWRM policy and water law, and has developed a comprehensive legal instrument, the WFD, that mandates river basin planning and then goes beyond IWRM to bring in a goal of 'good ecological status' for all of Europe's waters. It has also driven the environmental law, and water law, of most of its Member States since the 1970s, so it seems a reasonable place to start the substantive analysis. The WFD has been implemented separately in England and in Scotland. In Scotland there was previously no comprehensive river basin planning; England had a system of catchment planning, but nonetheless significant changes have been required.

The WFD's purpose is 'to establish a framework for the protection of inland surface waters, transitional waters [estuaries], coastal waters [up to one nautical mile from the low tide mark], and groundwater...'[29] It has been described as the 'third wave' of EU water law;[30] the first wave being the water pollution directives of the 1970s, and the second being the structural measures on water services and water quality of the 1980s and 1990s. The WFD was also a Fifth Environmental Action Programme directive,[31] with the emphasis not just on pollution control but on the sustainable management of the water resource.

It addresses water pollution and water supply, surface and ground waters, inland and coastal waters, water quality and water quantity. For water quality, the primary determinant is ecological quality, and secondary to that, chemical quality. The overarching objective of the Directive is to achieve 'good' ecological status, as defined; the definition is an ongoing process at EU and state level. It takes a dual or combined approach to pollution control, using both emission limit values and environmental quality standards. Essentially it is an IWRM system, which goes further by establishing a classification system for surface waters based on their ecological health, and by determining that, as a first principle, there should be a status target of 'good' (in a five-fold classification: high, good, medium, poor and bad). Good status is measured by identifying the capacity of a water body to support an acceptable range of ecosystems, by using a reference water body of an appropriate ecotype.[32]

The primary management tool of the WFD is the River Basin Management Plan (RBMP).[33] As noted above, river basins are identified and assigned to river basin districts (RBDs); these may be combinations of basins, but the hydrological unit may not be split.[34] Coastal waters and groundwater should be allocated to the most appropriate basin. Where a basin crosses state boundaries, international RBDs must be created; where the boundary is with a non-Member State, the Member State should seek to obtain their cooperation.[35]

As with any IWRM system, in order to produce the plans it is necessary to undertake a comprehensive review and analysis of the water resources. For the first round of plans, this involved mapping the water bodies, assessing their environmental condition, and undertaking an economic analysis and a review of human impacts.[36] These all formed part of the early 'characterisation' reports.[37] The Directive also makes detailed provision for different types of monitoring,[38] and a complex system of intercalibration, to enable comparisons of data across Member States. From this analysis a Programme of Measures would be drawn up for each water body[39] with the general objective of achieving 'good' ecological status.[40] Where a water body is 'heavily modified', i.e. its character has been substantially changed as a result of physical human activity, or it is an artificial water body, then the requirement is to achieve good ecological 'potential'. Good chemical status (i.e. compliance with all relevant chemical standards) must still be met. Ideally, these

---

[27] Convention on Access to Information, Public Participation in Decision-making and Access to Justice in Environmental Matters (Aarhus) (1999). Although this is a UN/ECE Convention and therefore neither South Africa nor Australia are signatories, it is widely referred to in international literature; and they too make corresponding provision, as befits signatories to the Rio Declaration and engagement with the global sustainable development agenda.

[28] See, e.g., UN (1992a) para.18.9(c); Dublin Statement (1992) Principle 2; and Chapter 1.

[29] WFD Art.1.   [30] Kallis and Nijkamp (2000); see also Kaika (2003).

[31] European Commission (1993). The 5th Programme was produced subsequent to the Rio Earth Summit, and as a policy document took forward the sustainable development agenda within the EU.

[32] E.g., a pristine highland river within the Scottish uplands, or an unmanaged chalk stream in the English lowlands; WFD Annexes II, V and XI. Reference bodies will be of 'high' quality and support the entire range of ecosystems found in a body of that type; and see further Chapter 4.

[33] WFD Art.13, Annex VII.   [34] WFD Art.3.   [35] WFD Art.13.

[36] WFD Art.5, Annex III.   [37] WFD Art.5, Annex II.

[38] WFD Art.8, Annex V; the WFD provides for surveillance, operational and investigative monitoring.

[39] WFD Art.11, Annex VI.   [40] WFD Art.4, Annex V.

measures would be implemented and the objective achieved by 2015, but failing that, the WFD envisages two more rounds of RBMPs, taking the whole process to 2027.

There is also provision for extensions and exemptions. The deadline of 2015 (the end of the first RBMP period) may be extended by two further updates of the plan (another 12 years) as long as there is no further deterioration and where one of the following conditions is met: technical feasibility, disproportionate expense or natural conditions.[41] Such extensions and the reasons must be set out in the plan, providing transparency and making explicit trade-offs between water quality and economic or social activity. Member States may set 'less stringent environmental objectives' (i.e. permanent exemptions) where water bodies are so affected by human activity, or natural conditions are such that the objective of good quality cannot be met, and all of the following conditions are satisfied: the environmental and socio-economic needs served cannot be achieved by other means that are a significantly better environmental option not entailing disproportionate costs; the highest possible status is achieved; and there is no further deterioration.[42] Again, these trade-offs must be set out in the plan. There is also provision for temporary deterioration by *force majeure,* and deterioration as a result of new sustainable human development activities, as defences to a failure to achieve the objectives.

These extensions and exemptions are sometimes referred to as 'alternative objectives' (though this is not a Directive term and may give a misleading impression). The complexity is increased by the provision for artificial and heavily modified water bodies. Only where these structures or modifications predated the Directive process do the special rules on ecological potential apply. Where a new artificial or heavily modified structure is proposed, such as a new dam, the applicable provision allowing a failure to achieve good status will centre around 'sustainable human development activities', for reasons of 'overriding public interest' and environmental and social benefit.[43] With hindsight, the additional complexity of 'ecological potential' for historic modifications may have been an unnecessary addition; it might have been easier to accept that such water bodies would be unlikely to achieve good status and classify them within the general system.[44]

The full detail of the WFD is beyond the scope of this book. Of particular note may be 'protected areas',[45] where water bodies specially managed under other Community legislation, including drinking water, recreational waters, or waters needed for species or habitat protection, must meet the higher standards laid down in other Community legislation. There is a requirement for a register of water abstraction and a system of abstraction control,[46] and a requirement to 'take account of the principle of cost recovery' in setting charges for water services.[47]

This last has proved one of the most contentious issues under the WFD. 'Water services' is defined, very broadly, to include all uses of the water environment, not restricted to urban water supply. If this includes agricultural water, then very few states comply. One of the most significant criticisms made of the first round of RBMPs was that most states did not implement full cost recovery outwith urban water services, and at the time of writing this question has been referred to the European Court of Justice.[48]

At the time of writing, Member States are preparing for the second round of RBMPs, to be finalised in 2015. Some preliminary work in the first planning period, such as the characterisation report and economic analysis, do not need to be repeated, but the broad participation requirements still apply.

2.4.1 The EU WFD participation requirements

Under the WFD, Member States should 'encourage the active involvement of interested parties' in the RBMP process.[49] Hence there is an active duty, but only towards 'interested parties'. Consultations should be made available to 'the public'. The process includes production of a timetable and work programme, three years before the planning period begins; an overview of water management issues, two years before; and a draft Plan, one year before. All documents must have a minimum six months' consultation. In both Scotland and England, the legislation provides a list of the 'interested parties' who will be directly consulted by the regulator, whilst policy documents expand both the mechanisms for engagement and how these apply to different groups of stakeholders. There is guidance from the Commission through the Common Implementation Strategy, indicating the priority given to participation as part of IWRM.[50] The WFD requires that national competent authorities will release any background documents or preparatory information on request, and that is a very useful provision that could and should be replicated.

2.5 SCOTLAND

Scotland is a small country with an abundant water resource, and therefore many might feel it is not a place from which lessons can be learned. It is perhaps therefore presumptuous to place

---

[41] WFD Art.4(4).   [42] WFD Art.4(5).   [43] WFD Art.4(7).

[44] For discussion of the issues around heavily modified water bodies, see European Commission (2012) Section 8.

[45] WFD Art.6 and Annex IV.

[46] WFD Art.11; this was relevant in Scotland where no such comprehensive system existed; see also Chapter 3.

[47] WFD Art.9; see also Chapter 5.

[48] *European Commission v Germany* (Case C-525/12). For this, and other referrals to the Court over the WFD, see European Commission (2012a) Section 6.

[49] WFD Art.14.

[50] European Commission (2002); see also HarmoniCOP (2005).

Scotland first in any substantive analysis of water law, and other chapters will not do so. However, this book is being written in Scotland, and Scotland took a very proactive and forward-looking approach to implementation of the WFD, so there is perhaps a justification beyond partiality. Scotland began the WFD process without either a history of statutory catchment management or any comprehensive abstraction controls, and an outdated system of pollution control. However, the establishment of the Scottish Parliament, with devolved powers for environment and water, enabled law reform in areas previously restricted by lack of Parliamentary time at Westminster;[51] and there was much enthusiasm for reform both in water resources and in water services; it might be described as a happy confluence, for water law. The WFD was implemented by primary legislation under the WEWS Act, not secondary regulation as in England. The Act also established a framework for a comprehensive review of water pollution control in Scotland, enabling a new system of combined water use licences to manage discharges, abstractions, impoundments and river engineering in regulations (below). A similar approach is seen in South Africa, which also uses integrated water use licences for discharges and other uses of the resource; there are arguments for and against this, but it does allow a policy focus on water.

In Scotland, the competent authority for the WFD is the principal environmental regulator, the Scottish Environment Protection Agency (SEPA), set up in 1995.[52] SEPA's functions regarding the water environment were relatively limited, including the control of water pollution but not catchment planning, and the provision of flood warning systems but not responsibility for flood defence;[53] and, separately, a limited scheme for control of abstractions for commercial irrigation only.[54] SEPA's primary duty was only 'to promote the cleanliness' of water resources and 'to conserve [them] so far as practicable'.[55] Secondary to this were duties to promote conservation and enhancement of beauty and amenity, and conserve aquatic flora and fauna, but (unlike the corresponding provision in England) there was no broad duty on the conservation or use of water resources, nor any mention of recreational use. The WEWS Act was a major step forward in many ways.

2.5.1 The legislative framework

The WEWS begins with a statement of its general purpose, which restates much of Art.1 of the WFD. The primary purpose therefore is 'protecting the water environment', which in turn includes preventing deterioration of ecosystems, promoting

sustainable water use, and protecting and improving the water environment. Key aims include providing a sufficient supply of water 'as needed for sustainable, balanced and equitable use'.[56] There are general duties on the Ministers, SEPA and other 'responsible authorities'[57] including a duty on the Ministers and SEPA to 'exercise their functions under the relevant enactments so as to secure compliance with the requirements of the Directive'.[58] In addition, all responsible authorities must exercise their designated functions in order to secure compliance with the WFD. These are significantly broader than the duties in the English regulations.

There is also a 'sustainable development' duty, and a duty to promote sustainable flood management, both of which took up much time in the Parliamentary Committee that scrutinised the Bill.[59] The sustainable development duty is couched in very general terms, requiring the authorities to 'act in the way best calculated to contribute to the achievement of sustainable development'. Importantly for the links with other planning processes, there is an obligation on all the authorities to 'so far as practicable, adopt an integrated approach by cooperating with each other with a view to coordinating the exercise of their respective functions'. Although not expressly stated, it is arguable that these provisions could give the RBMP primacy insofar as those other plans impact on the water environment, although that is not the view of the Scottish Government.[60]

As noted above, definitions are for the most part transposed directly from the WFD, with the exception of the inclusion of 'wetlands' in the 'water environment'.[61] This brings wetlands within the control regime for regulating water use and was one of the reasons why the Scottish Government was accused of 'gold-plating' this Directive, not least by the Parliamentary Finance Committee.[62]

Part 1 of WEWS then sets out a framework for the implementation of the WFD through the RBMPs. The RBDs are to be established by Ministerial Order.[63] SEPA must undertake the characterisation reports for each RBD, including the economic analysis and review of human impacts,[64] and then a report on the significant water management issues in each basin, which in turn will feed into the draft RBMPs. Water bodies used for drinking

---

[51] Hendry (2003).   [52] Environment Act 1995 c.25 Part I.
[53] Environment Act 1995 s.21.
[54] Natural Heritage (Scotland) Act 1991 c.28.
[55] Environment Act 1995 s.34.

[56] WEWS s.1.   [57] WEWS s.2.
[58] The authorities and the enactments are specified by order; currently, the Water Environment (Relevant Enactments and Designation of Responsible Authorities and Functions) (Scotland) Order SSI 2011/368.
[59] Scottish Parliament Transport and Environment Committee (2002).
[60] Scottish Executive (2006).   [61] WEWS s.3.
[62] Scottish Parliament Finance Committee (2002). 'Gold-plating' indicates going beyond the requirements of the Directive and is generally perceived as a negative attribute in the UK, although the Committee here was more concerned with general costs of implementation and the increased burden likely to be faced by the various authorities, and by agriculture and industry regarding abstraction costs.
[63] WEWS s.4.   [64] WEWS s.5.

water abstraction must be identified[65] and these are 'protected areas',[66] along with conservation sites under EU law.[67] There is no conceptual reason why sites with national protective designations should not be noted within the RBMP merely because it is not a Directive requirement, and this would be desirable. The protected areas, as required by the WFD, must be mapped and the maps are available on SEPA's website; visual displays of information are useful to authorities, water users and the general public, and should be encouraged, if not required.

The Act then provides for the monitoring programme required by the WFD, and the setting of environmental objectives and programmes of measures,[68] and outlines the RBMP process.[69] The Draft RBMPs will be submitted to the Ministers and should include summaries of the characterisation process, the pressures and impacts, protected areas, monitoring, objectives and the programme of measures, along with a non-technical summary.[70] There are requirements for consultation and publicity,[71] and the plan must be approved by the Ministers[72] and reviewed every six years.[73]

2.5.2 Controlled activities

The WEWS Act enables regulation of 'controlled activities', i.e. abstractions, impoundments, discharges and river engineering.[74] The Controlled Activities Regulations (CAR)[75] then brought in a comprehensive and integrated system to authorise all these activities. The CAR uses a proportionate three-tier system based on risk assessment and comprising general binding rules (GBRs) for the smallest scale and least damaging uses; registration for activities that may have cumulative impacts and should be identified to the regulator; and full licences for the most potentially damaging operations. These rules enabled the complete revision of historic law on abstractions and discharges, and as such will be considered in Chapters 3 and 4; at the time of writing they are being used in Scotland as the basis of further reform of environmental law.

2.5.3 Water resource planning by the water services provider

Scottish Water (SW) is the public water services provider, serving most of Scotland, and has recently started to develop and publish 25 year water resource plans.[76] The company is a major

abstractor and discharger and an important stakeholder in managing Scotland's water environment, and has a significant role to play in achieving the objectives of the WFD. The Water Resource Plan addresses security of supply nationally and within water supply zones, and will be updated in each new price review period (six-yearly). Formal water resource planning by the service provider is relatively new in Scotland, but well established in England, where it will be addressed more fully. In 2013, for the first time, as part of the business planning process (Chapter 5) SW has also produced a 25 year 'Strategic Projection' within which the six-year business plan will be contextualised.[77]

2.5.4 The river basin management plans

Two RBDs are mainly or wholly in Scotland: the Scotland RBD, covering much of the mainland and the Northern and Western Isles, and managed by SEPA; and the Solway Tweed RBD, running across the south of Scotland and into the north of England, and managed by SEPA working with the English Environment Agency. The first RBMPs were published in 2009 as required and indicated that some 65% of water bodies in the Scotland RBD, and around 50% in the Solway Tweed, were at 'good' status or better.[78] As the WFD process lasts till 2027, in Scotland targets for improvements are also set in the RBMPs, and updated in each planning cycle, with further programmes of measures for each water body not reaching good status, and exemptions or extensions if required. The targets for 2021 and especially 2027 were significantly increased between the submission of the draft Plans in 2008 and the final versions in 2009, and the final targets were for 97% of waters in the Scotland RBD, and 92% of those on the Solway Tweed, to reach good status. This was especially as a result of a comment by the Scottish Government, which considered the draft Plans to be lacking in ambition.[79] There is a specific requirement in WEWS that SEPA must, when submitting the Plan for approval by the Ministers, also provide a summary of any representations made to the draft Plan and amendments made as a result. This type of provision is very valuable in terms of facilitating accountability and access to justice, and is recommended in other jurisdictions and other aspects of environmental management.

2.5.5 Participation

As noted, the WFD sets a timetable for consultation on key documents, to be made available to the general public, and a requirement that Member States 'encourage the active involvement of all interested parties', supported by extensive guidance.

---

[65] WEWS s.6.
[66] WEWS s.7; Water Environment (Register of Protected Areas) (Scotland) Regulations 2004 SSI 2004/516.
[67] Principally the Natura2000 system under Directives 2009/147/EC (Birds) and 1992/43/EEC (Habitats), along with designated fish and shellfish areas, Directives 2006/113/EC and 2006/44/EC.
[68] WEWS ss.8–9.   [69] WEWS ss.10–17.
[70] WEWS Sch.1; WFD Annex VII.   [71] WEWS s.11.
[72] WEWS ss.12–13.   [73] WEWS s.14.   [74] WEWS s.20.
[75] Water Environment (Controlled Activities) (Scotland) Regulations SSI 2005/348 (CAR 2005), now SSI 2011/209 (CAR).
[76] Scottish Water (2009).

[77] Scottish Water (2013); and see Chapter 5 on business planning.
[78] Scottish Government (2009), EA/Scottish Government (2009).
[79] See SEPA (2009, 2009a).

In Scotland, WEWS lists certain stakeholders who must be consulted by SEPA, including local authorities, SW, nature conservation bodies, fisheries bodies, business interests and environmental groups.[80] WEWS also provides for advisory groups. Currently, there is a National Advisory Group with a wide membership,[81] and there are Area Advisory Groups for the eight sub-basins in the Scotland district and the two sub-basins in the Solway Tweed.[82] The advisory groups were drawn from the list of core stakeholders in WEWS. In the first planning round, each of these sub-basins also had its own sub-basin plan, and there were also advisory forums to allow involvement from the wider public.

Currently, SEPA is moving into the preparation of the second round of RBMPs, and some changes will be made to reflect lessons learned, and to streamline the process. At the time of writing, SEPA has consulted on the work programme,[83] and on significant water management issues.[84] One important intention is to focus more on small-scale catchment working, taking advantage of the expertise that has developed through the Area Advisory Groups and similar bodies.[85] This question of scale is critical to IWRM implementation, ensuring that planning frameworks operate at a scale large enough to be efficient, whilst implementation on the ground is at a level that enables meaningful involvement of users. This is also relevant in England, where implementation of the WFD has been rather different.

As well as the ongoing RBMP process, the Scottish Government and Parliament have continued to be active in water, with new legislation affecting large-scale abstractions, as well as ongoing reforms to water services; see Chapters 3 and 5.

2.6 ENGLAND

The current legislative provision for water resource management in England is complex. The Water Resources Act 1991 (WRA)[86] is the principal legislation for abstractions and impoundments, albeit much amended, especially by the Water Act 2003 (WA2003).[87] The functions and powers relevant to IWRM are now mainly contained in the Environment Act 1995, which created the Environment Agency (EA, the Agency), the principal pollution control agency in England. Water-related functions

exercised by the EA include water resource management, water pollution, flood defence, land drainage, fisheries regulation, and some responsibilities for navigation, harbours and conservation.[88] Specific water resource management duties include the conservation and enhancement of inland and coastal waters, conservation of aquatic flora and fauna, the recreational use of waters, and the conservation, redistribution, augmentation and proper use of water resources.[89] The EA's powers over water, reflecting its predecessor body, were accordingly broader than those of SEPA prior to the WFD and WEWS. The Government sets the wider policy context for both water resources and water services, has powers of direction (over the EA and other public bodies), and issues guidance to the EA and water companies.

2.6.1 Water resource planning prior to the Water Framework Directive

The bare enactments give little indication of the extent of water resource management in England already existing prior to the implementation of the WFD. As an early step in the WFD process, the EA produced a consultation strategy document[90] which set out *inter alia* the multiplicity of existing plans, and identified 10 groups of plans where the EA took the lead, including national and regional water resources strategies. Further work in the WFD process identified five sets of planning processes not led by the EA but which should be linked to the RBMP process,[91] including land use planning, biodiversity, and coastal and marine planning. Howarth suggested that the WFD would be a valuable impetus for rationalisation of these many documents;[92] but that has not really materialised.

2.6.2 National water resources strategy and plans

The national water resources strategy[93] has a long time frame, till 2050, and is supplemented by national and regional action plans. Its title is 'Water for People and the Environment' and it is accordingly concerned with managing supply – 'resource development' – and demand – the 'twin track' approach – and increasing resilience in the context of pressures, including population growth and climate change. The strategy recognises the links with both the WFD process and water company water resource plans, and the role of diverse policy tools from water pricing to reforms of abstraction licences. The focus is on abstraction and water use, whilst recognising the links to water quality and catchment management processes (within and outwith the WFD) and the roles of the water companies. The national action plan sets out short-, medium- and long-term activities (for the EA and others) to achieve the strategy;[94] the regional action plans

---

[80] WEWS s.11.

[81] 'SEPA National Advisory Group membership and papers' see http://www.sepa.org.uk/water/river_basin_planning/national_advisory_group.aspx.

[82] 'SEPA Area Advisory Groups membership and papers' see http://www.sepa.org.uk/water/river_basin_planning/area_advisory_groups.aspx.

[83] SEPA (2012), SEPA (2013).   [84] SEPA (2013a).

[85] Especially, the Fish and Fisheries Advisory Group and the Diffuse Pollution Management Advisory Group, considered further in Chapter 4.

[86] Water Resources Act 1991 c.57 (WRA).

[87] Water Act 2003 c.37 (WA2003).

[88] WRA s.2.   [89] WRA s.6.   [90] EA (2005).   [91] EA (2006).

[92] Howarth (2005).   [93] EA (2009).   [94] EA (2010).

also include data and analysis specific to that region.[95] The seven EA regions do not directly correspond to the boundaries of the water services companies, though they have some similarities, but they are the basis for the English RBDs. The EA also produces national and regional drought plans, and these are concerned with temporary reallocations of water, to ensure the primacy of public supply and to manage the ecological, social and economic consequences of the reallocation.[96]

2.6.3 Water company water resource plans

Currently, there are 10 regional water and sewerage companies and, in addition, there are local water supply companies and licensed suppliers. The complex structures will be considered in Chapter 5, but, in all the UK jurisdictions, water services providers are major abstractors of water and are expected to make long-term plans for water use. The water resource management plans have a 25 year horizon, but are reviewed five-yearly as part of the periodic price review. They were placed on a statutory footing under the WA2003,[97] making public consultation on the draft plans mandatory; this could be seen as part of a much greater emphasis on customers' views and wishes that has emerged in the regulation of the sector and will also be considered in Chapter 5. The national strategy and other guidance, from the EA, the Government, and the industry regulator, all set the context for these plans.[98] The Water Bill currently before Parliament will require these plans to address resilience in future.[99]

The water resource plans set out the anticipated levels of consumption, leakage and supply–demand balance for the next five years, for subsequent price reviews up to 25 years, and some future-casting beyond that. Clearly different water services providers face different challenges – in Anglian Water, in the southeast, where water is very scarce and where population is expected to increase significantly, the pressure to reduce household consumption is very strong. Anglian Water is expected to make significant 'sustainability savings' in terms of their abstraction regime. Part of their response has been to work more closely with the other smaller water supply companies in the region, to implement more water trading and bulk transfers, in line with Government and EA wishes.[100] In the northwest by comparison, United Utilities are still keen to reduce consumption, but unsurprisingly there is more focus on catchment management, biodiversity protection and the

impacts of flooding.[101] The impacts of climate change will be different, and water companies in the north are perhaps more likely to be selling water than buying it in any new trading regime – although at present most bulk trades are within regions rather than cross-country (see Chapter 3). Along with the water company business plans, these plans identify investments over the planning period. The water companies also produce statutory drought plans (Chapter 5).

2.6.4 Catchment abstraction management strategies

Catchment abstraction management strategies (CAMS) were introduced by the EA in order to improve the transparency and sustainability of the abstraction licensing process; hence they are also relevant to Chapter 3, but the planning element will be considered here. CAMS is a non-statutory system which enables a better appraisal of water needs, including environmental needs. It was established in 2002, and reviewed to take account of changes to the licensing regime in the WA2003, and the introduction of the WFD.[102] CAMS should be a transparent method for balancing different needs of different water users, including environmental needs, by indicating where there is water available to abstract; it plays an important role in implementing the WFD.

CAMS is a staged process, beginning with a resource assessment that feeds into detailed catchment-based licensing strategies, and then potentially a reduction in water abstracted via the 'Restoring Sustainable Abstraction' programme.[103] Assessments consider what water is available for abstraction at different flows, using the Environmental Flow Indicator, which is a percentage deviation from the river's natural flow. CAMS then provides information to inform not just decisions on new licence applications, but also decisions on renewing or varying existing licences.[104] Conditions may be placed on licence-holders in certain flow conditions ('hands-off flow' and 'hands-off level' conditions). The assessment may indicate a need for reductions in abstractions to ensure that these remain sustainable, especially for protected habitats; it also links into the wider determination of ecological status under the WFD, of which adequate flow is one component. The Restoring Sustainable Abstraction process is carried out via CAMS for most abstractors, although for water companies there is a separate national process establishing targets and objectives for sustainability reductions, to ensure that their actions support and are coordinated with the WFD programmes of measures.[105]

---

[95] See, e.g., EA (2009a).

[96] See generally 'Environment Agency Drought Planning' http://www. environment-agency.gov.uk/homeandleisure/drought/31771.aspx.

[97] Water Act 2003 s.62.

[98] See especially DEFRA (2012); and EA/OFWAT/DEFRA/Welsh Government (2012).

[99] Water Bill (UK) 2013 Bill No.82 s.27.   [100] Anglian Water (2014)

[101] United Utilities (2013).   [102] EA (2013).   [103] EA (2013).

[104] The licensing process, and the circumstances in which a licence can be withdrawn or varied, will be addressed in Chapter 3.

[105] EA (2013a).

2.6.5   Implementation of the Water Framework Directive

Despite the plethora of existing plans, the recognition of the desirability of rationalising the same, and the passage through Parliament of the WA2003 just at the time of implementing the WFD,[106] nonetheless the implementation of the WFD in England essentially took place as a stand-alone process – in some ways the antithesis of a true IWRM approach. The Directive was implemented by Regulations (the WFD Regulations)[107] that establish the EA as the lead authority and give the Secretary of State responsibility for approving the RBMP. The Secretary of State can issue directions to the EA and other public bodies, and the EA and the Secretary of State must exercise 'their relevant functions so as to secure compliance with the requirements of the Directive'.[108]

The Regulations mandate the WFD process, including the initial characterisation reports, the monitoring network, the identification of significant water management issues, and the draft and final RBMPs. Compared to the extensive reforms undertaken in Scotland, the transposition in England was minimalist. Thus, for example, whereas in Scotland supplementary plans are mandatory, in England they are not. Definitions in the schedules only apply for WFD purposes – they did not repeal pre-existing definitions of elements of the water environment in other legislation, and this left different definitions, for example of coastal waters, transitional waters, groundwater and ground waters, which is surely a recipe for confusion.

2.6.6   River basin management plans and planning

There are 11 designated RBDs in England and Wales; two of these cross the border with Wales (one is solely in Wales) and two cross the border with Scotland – Northumbria and the Solway Tweed. The constituting regulations for the latter are made by the UK Parliament,[109] and there is also extensive guidance on the relationship between SEPA and the EA in the production of the RBMP for the Solway Tweed. The WFD process operates along with the wider range of planning and strategy documents discussed above. The EA has issued overarching policy that is relevant to the second round of RBMPs,[110] as well as specific consultations.

When the first RBMPs for England were issued, on average only 26% of water bodies achieved good (or high) status.[111] As in Scotland, morphology, diffuse pollution and over-abstraction are the main reasons why surface waters failed. Whilst it is recognised that England has a high population density and a legacy of industrial pollution, nonetheless this was a poor result. Furthermore, the Government seemed reluctant to introduce ambitious programmes for improvement, preferring to use the Article 4 extensions and exemptions. After the RBMPs were published, the Government was threatened with a judicial review by the WWF and the Angling Trust, on the basis that they had not adequately complied with the WFD.[112] In response to this, the Department for Environment, Food and Rural Affairs (DEFRA) announced a new approach to river basin planning that would address the scale gap and enable a better focus on catchment activities.[113] Some of these (such as the demonstration test catchments) have particular relevance to water quality and are also relevant to Chapter 4. The general approach, and attention to scale, is of interest in the context of IWRM.

2.6.7   Participation in river basin planning

As in Scotland, the WFD Regulations list persons and bodies that must be given the opportunity for active participation, and again they include a range of public bodies and non-governmental organisations; 'public bodies' include statutory undertakers.[114]

At national level, DEFRA has previously hosted a Stakeholder Forum, which met once or twice per year. The EA convenes a National Liaison Panel,[115] and at RBD level there are also Liaison Panels of 12 to 15 people, representing the most significant regulatory, service provider and interest groups. Given the size and scale of the RBDs in England, 10 panels of 15 people does not amount to significant stakeholder engagement, regardless of intention or willingness to cascade information. Thus the new focus on smaller-scale local and catchment initiatives will be very important in the second round of plans.

At the time of writing, the EA has consulted at national level regarding stakeholder engagement in the second round of plans, and has issued a response to the replies it received.[116] It is currently consulting at national and RBD level, around the most significant water management issues,[117] and the resulting policy choices. This will be followed by the draft RBMPs in 2014. The key challenges for the second round are identified as abstractions/flows, chemicals, faecal and sanitary pollutants, invasive non-native species, nitrates, phosphates and physical modification. This is a long list of challenges and it is unclear how, in

---

[106] The Government rejected requests to use the WA2003 to transpose the WFD; see, e.g., at second reading, Hansard HL [Vol.654] Col.972.

[107] The Water Environment (Water Framework Directive) (England and Wales) Regulations SI 2003/3242 (WFD Regulations).

[108] WFD Regulations Reg.3.

[109] Water Environment (Water Framework Directive) (Northumbria RBD) Regulations 2003 SI 2003/3245, Water Environment (Water Framework Directive) (Solway Tweed RBD) Regulations 2004 SI 2004/99.

[110] EA (2013b).

[111] EA (2009b).   [112] WWF (2011); Angling Trust (2011).

[113] DEFRA (2013).   [114] WFD Regulations Reg.2.

[115] 'Environment Agency National Liaison Panel' see http://www.environment-agency.gov.uk/research/planning/33114.aspx.

[116] EA (2012).   [117] EA (2013c).

difficult financial circumstances, the Government will be able to meet them, even with the active involvement of a wide range of stakeholders. It is also likely that the European Commission will expect a much more systematic and comprehensive approach in the next iteration.

2.7   AUSTRALIA

As noted in Chapter 1, constitutional responsibility for water lies with the states, but federal Australia has responsibility for trade. In the last 20 years the Commonwealth has used its trading powers, and a combination of legal, financial and political means, to drive forward a water reform agenda within the states that has focused on trade and efficiency. Much of this will be relevant to later chapters, but given the complexity of the regimes in Australia, a brief introduction here may be useful.

In 1992 the Council of Australian Governments (COAG) produced a National Strategy for Ecologically Sustainable Development,[118] responding to Agenda 21. 'Principles of ecologically sustainable development' are commonly found in relevant legislation in the Australian states. The Strategy identifies IWRM as a key objective. In 1994, COAG issued a water policy identifying the need for greater efficiency, water trading and full cost recovery.[119] In 2004, the National Water Initiative (NWI) sought to extend market reforms, provide for environmental water and clarify how risk should be shared.[120] In turn, the NWI led to the Water Act 2007,[121] affecting the Murray–Darling Basin.

Unsurprisingly in the Australian context, this reform process – and most water planning – has essentially been about management of water quantity. Whilst there have also been numerous initiatives in water quality, the two have not always been clearly integrated, though this is much less true since the Water Act 2007 and the development of the Murray–Darling Basin Plan (below); at Commonwealth and state level, this legislation makes it clear that ecology is a key driver for management of quantity. The National Water Quality Management Strategy (NWQMS)[122] provides *inter alia* water quality guidelines for specific uses, including ecological quality. In addition there has been a series of non-statutory planning mechanisms for natural resources generally, focused on agricultural land and currently under the heading of 'Caring for our Country'.[123] Operating on a regional basis, the regions for this follow catchment boundaries.

The lengthy drought at the start of the new millennium was a key factor in the most recent reforms, but the political context is also important, both the relations between the Commonwealth and states, and between different states, especially in the context of the Murray–Darling.[124]

2.7.1   The Murray–Darling Basin Authority and the Water Act 2007

The Murray–Darling Basin is Australia's largest river system, covering over 1 million square kilometres and supplying water and other services for up to 3.3 million people.[125] It is heavily developed and under threat from over-exploitation. The headwaters of the River Murray are in Queensland, and those of the River Darling in Victoria; the mouth of the basin is in South Australia. It also drains land in New South Wales, Victoria and the Australian Capital Territory. For the best part of a hundred years there have been inter-state agreements to manage the Murray and then the Murray–Darling.

The Murray–Darling Basin Authority was established under the Water Act 2007 and is a statutory body.[126] The Act itself is detailed and complex and there has been much academic commentary, from a variety of disciplinary perspectives.[127] This section will identify the key features, look at the water planning mechanisms, and consider the issues around stakeholder engagement. The Murray–Darling is a 'wicked' problem, with multi-layered administration and governance.[128]

The Act makes read-down provision to ensure the constitutionality of Commonwealth activities, and interpretative provision to enable the concurrent implementation of Commonwealth and state law.[129] It also utilises Australia's commitments under international law to justify some of the activities of the Commonwealth and the Authority.[130] It provides for a Basin Plan, which must include sustainable diversion limits, trading rules, and sub-plans on water quality and salinity, and on environmental watering.[131] The Plan has been published and adopted by the Minister.[132] The Plan is a legislative instrument,[133] as will be the Water Resources Plans that must subsequently be made by

---

[118]   COAG (1992).   [119]   COAG (1994), see also Chapters 3 and 5.
[120]   COAG (2004).
[121]   Water (Cwlth) Act 2007 No.137 (Water Act 2007) as amended, especially by Water Amendment (Cwlth) Act 2008 No.139.
[122]   'NWQMS' see http://www.environment.gov.au/water/policy-programs/ nwqms/index.html and Chapter 4.
[123]   'Caring for our Country' see http://www.nrm.gov.au/index.html.

[124]   See, for some of that political history, Connell (2011).
[125]   'Murray–Darling Basin' see generally http://www.mdba.gov.au/.
[126]   Water Act 2007 Parts 9 and 10; and see Sch.1 for the text of the Murray–Darling Basin Agreement between the States.
[127]   See, e.g., Connell and Grafton (Eds.) (2011), and see further Chapter 3.
[128]   See, e.g., Wallace and Ison (2011), Garrick *et al.* (2012).
[129]   Water Act 2007 ss.9–9A; ss.36–37; ss.60–61; and Part 11A, on interaction with state laws.
[130]   Especially, commitments under the Convention on Wetlands of International Importance (Ramsar) (1971) and the Convention on Biological Diversity (CBD) (1992); see, e.g., s.20, s.21(3).
[131]   Water Act ss.20–27, ss.28–32; and Chapter 3.
[132]   Commonwealth of Australia Water Act 2007 Basin Plan 2012.
[133]   Water Act 2007 s.33.

the Basin States.[134] 'Critical human water needs' are given the highest priority for water use,[135] and there are special provisions for water sharing in the event that these needs are likely not to be met. Perhaps more interesting though is the focus on ecology, including but not restricted to ecosystems protected under international law, especially the Ramsar Convention. It has been suggested that the international framework has been used by the Federal Government not just to alter domestic law, but specifically to change the constitutional relationship between Commonwealth and states.[136] The environmental watering plan is designed to restore water to the environment, and the diversion limits and trading rules are designed to provide water for that purpose, whilst the water quality and salinity plan, and associated objectives and targets, are intended to ensure that the resultant water quality is adequate to protect those ecosystems. There is no overall mandatory target for ecological quality as there is in the EU, though in some ways the Australian model is more advanced, with its emphasis on ecosystem services and functions.[137]

2.7.2 Participation in the Basin Plan

The Act sets out a timetable for consultation with the basin states and other stakeholders, in preparation of the draft Plan, giving minimum periods for consultation and also requiring the Authority to explain how it took account of comments received.[138] The brief outline in the Act sets out the process but inevitably does not reflect either the detail or the heated debate engendered.

In October 2010, the Authority issued a pre-consultation 'Guide' document, suggesting that between 3000 and 4000 GL should be restored to the environment, and also suggesting that to ensure environmental health of the system up to 7000 GL might be required. The Guide document attracted widespread public criticism from irrigation communities,[139] and a response from the Government Solicitor indicating that equal weight should be given to the economic, social and environmental factors; the Chair of the Authority subsequently resigned, and a Parliamentary inquiry was held and reported in spring 2011.[140] The formal consultation on the draft Plan began in November 2011 and 12,000 submissions were received;[141] here the proposal was to

buy back 2750 GL, but also allow an extra 2600 GL of groundwater extractions. Criticism continued from environmental groups and at least some parts of the science community, including concern that the new groundwater allocations had not been properly modelled.[142] The Government issued formal advice on economic and social factors in May 2012,[143] and a second draft Plan was submitted to the Ministers in August 2012; the headline numbers remained unchanged when the Plan was finally adopted.

It is hard to be sure what lessons to take from this. One obvious problem was the perception that the Guide document was relatively final, as opposed to a very early stage of engagement. Another might be that advice on the socio-economic factors should have been issued at a much earlier stage. One analysis suggests that some 6000 jobs were lost as a result of the drought, and only around 500 as a result of the water buyback;[144] another suggested that reductions of up to 4400 GL would produce only a moderate reduction in profits basin-wide, albeit widely varied across catchments.[145] Both these analyses suggest the process could have been better explained to stakeholders. Nature had a role to play: several years of extreme drought had softened attitudes to water restrictions, but when the drought had broken that driver was much weaker. It is possible that in such a controversial process agreement was never going to be reached, and arguable that the Government should have made hard decisions earlier and not sought consensus; but, given that the whole rationale of the 'soft' aspects of IWRM is around consensus-building, and given the extensive resources available in a country like Australia, the early stages of stakeholder engagement seem to have been mismanaged.

2.8 QUEENSLAND

In Queensland, the QWA is the principal legislation for statutory water resource planning, along with a Water Regulation.[146] Chapter 2 of the QWA addresses allocation and sustainable management. Water Resource Plans (WRPs) determine how much water is available for allocation, and subsidiary Resource Operations Plans (ROPs) grant finalised water allocations which are then tradable. Where infrastructure is required as part of the allocation then this is approved in the form of a Resource Operations Licence (ROL); where the ROP is not finalised, then Interim ROLs will be granted. Much of this, including the trading regime, will be addressed in Chapter 3. The planning elements of

[134] Water Act 2007 s.57.   [135] Water Act 2007 s.86A.

[136] Garrick et al. (2012).

[137] See especially Schedule 11 of the Basin Plan, setting target values, mainly for physico-chemical parameters, for specified water bodies and wetlands; and see further Chapter 4.

[138] Water Act 2007 ss.41–44, and see 'Steps in the Development of the Basin Plan' http://www.mdba.gov.au/what-we-do/basin-plan/development/steps-in-the-development-of-the-basin-plan.

[139] See, e.g., Taylor (2010).

[140] House of Representatives Standing Committee on Regional Australia (2011).

[141] 'Murray–Darling Basin Plan consultation documents' see http://www.mdba.gov.au/what-we-do/basin-plan/consultation.

[142] Wentworth Group (2012).

[143] Australian Government/Murray–Darling Basin Authority (2012).

[144] Wittwer (2011).   [145] Grafton and Jiang (2011).

[146] Water Regulation 2002 SL No.70, containing more detail, for example on licensing processes.

the WRPs, and to a lesser extent the ROPs, will be examined here, as will the non-statutory catchment management systems.

In addition, there have been several legislative changes in recent years, relevant mostly to bulk supply (Chapter 3) and the delivery of water services (Chapter 5). There is also specific legislation clarifying certain matters in relation to the Water Act 2007, for the Murray–Darling Basin.[147]

The institutional and departmental frameworks in Queensland have been subject to much change in recent years, and some of this is ongoing at the time of writing, since the 2012 state elections. Many aspects of water regulation, especially of water services (bulk and retail), are now under the control of the Department of Energy and Water Supply (DEWS), although water resource planning remains for now with the Department of Natural Resources and Mines (DNRM), and water quality, along with coastal zone management, is with the Department of Environment and Heritage Protection (DEHP). The Department of State Development, Infrastructure and Planning also has a role. The DEWS has recently consulted on a deregulation and innovation agenda, which is however specifically declared not to apply to water resource planning.[148]

2.8.1 Purposes and principles

The overall purpose of the QWA Chapter 2 is 'to advance sustainable management and efficient use of water and other resources by establishing a system for the planning, allocation and use of water'[149] and it then provides further definitions of 'sustainable management',[150] 'efficient use'[151] and also 'principles of ecologically sustainable development',[152] which themselves are part of sustainable management. Thus the policy context and aims are given legal force, and the policy purpose is clearly stated. Furthermore, there is a general requirement on all 'entities' with 'functions or powers' under the QWA to perform them in a way that advances the purposes of the chapter.[153] Protection of biodiversity and ecology is just as important to these provisions as social or economic benefit. Immediately sequential to these principles is provision that all rights in water vest in the state.[154]

2.8.2 Water resource plans

There are 22 WRP areas in Queensland, based on catchments, and a WRP is also prepared for the Great Artesian Basin.[155] The Minister has the power to prepare WRPs;[156] these are statutory instruments, and have (non-exclusive) purposes. They should define water availability; provide a framework for the sustainable management and taking of water; identify priorities and mechanisms for future water requirements; provide a framework for water allocations; and, 'where practicable', provide for reversing ecosystem degradation (but noting that sustainable management already includes prevention of further harm).

The WRPs are strategic, and have certain common features.[157] After setting out the Plan's area and the water covered, they establish outcomes for sustainable management, and provide *inter alia* for economic development, social and cultural values, water availability and quality, and ecological outcomes, both general and specific to the catchment. The WRPs then set performance indicators and objectives for environmental flows (for surface water) and water allocation security, and these are the constraints within which the detailed allocations can be established in the ROPs. Where relevant, WRPs may address groundwater or the use of overland flow. There are also area-specific requirements, especially for protection of sites of high ecological value (Chapter 4).

All WRPs provide for monitoring, and where there is infrastructure, i.e. dams and weirs ('supplemented water'), the holder of the ROL or Interim ROL is responsible for monitoring to ensure compliance with both volumetric limits and environmental flows. Where there is no infrastructure ('unsupplemented water', taken directly from the source) then volume controls tend to be specified by the rate of extraction and pump size, and the Department is the regulator. The WRP will specify matters that must be considered in relation to the ROP or other allocations. These might include the impact of infrastructure on stream flows and habitats, or the effect of rapid changes in water levels. There are powers to amend WRPs if they are not achieving their outcomes, or their objectives are no longer suitable.

2.8.3 Resource operation plans

The ROPs are much more detailed and specify the actual water allocations, the water sharing rules, and other technical requirements such as monitoring, in detail. These are not subordinate legislation but are produced by the Department, along with supporting documentation, including the Minister's annual reports on the WRP.[158] The primary purpose of an ROP is to provide finalised permanent allocations that can be traded. The ROPs identify the land and water to which the Plan applies, and the general principles and outcomes for sustainable management and allocation. They then set out the state monitoring and reporting requirements for water quantity and quality and ecosystems. There is an

[147] Water (Commonwealth Powers) (Qld) Act 2008 No.58.
[148] DEWS (2012); and see Chapter 5.   [149] QWA s.10.
[150] QWA s.10(2).   [151] QWA s.10(3).   [152] QWA s.11.
[153] QWA s.12.   [154] QWA s.19, and see Chapter 3.
[155] 'Queensland Water Resource Plans' see generally http://www.nrm.qld. gov.au/wrp/.
[156] QWA s.38.

[157] See, e.g., Water Resource (Condamine and Balonne) Plan 2004 SL No.151; Water Resource (Gold Coast) Plan 2006 SL No.321.
[158] See DNRM (2013).

overview of each of the water supply schemes, and the management areas for unsupplemented water. There is provision for the granting of new entitlements, either licences or allocations; for example, to non-riparians, to local governments, and, if permitted by the WRP, for the release of unallocated water in certain circumstances. The bulk of the ROP provides the detail for each supply scheme or management area. These include the specific allocations granted to individuals or companies; the rules for conversion to the same; the total volumes;[159] operating rules including infrastructure operation where appropriate; and the water sharing rules. In supply schemes the last are operated by the ROL holder and ensure supply for high priority users.

2.8.4 Implementation of the Murray–Darling Basin Plans

Some of the Queensland catchments are within the Murray–Darling Basin, including the Condamine–Balonne, Border Rivers, Moonie River and Warrego–Paroo–Bulloo–Nebine catchments. These catchments may need their WRPs revised to ensure compliance and it is likely that the duration of some of these WRPs will be extended to meet the planning periods under the Murray–Darling Plan. In addition, some other catchments may be managed under other interstate agreements, including the Great Artesian Basin, Border Rivers and Moonie River catchments.[160] For the upper Condamine, the only Queensland catchment where there will be buyback of groundwater under the Murray–Darling Basin Plan, a water management plan has been produced to take account of this and ensure that holders of groundwater are eligible to participate in the Commonwealth purchasing scheme.[161] The Water Act 2007 requires states to produce Water Resources Plans,[162] so this document is an Interim Water Resources Plan for that purpose, but, to avoid unnecessary confusion with Queensland statutory WRPs, it is being termed a water management plan. This brings together all the relevant planning instruments for Condamine, including a current moratorium on further groundwater allocations.

2.8.5 Natural resource management and non-statutory catchment activities

In Queensland, a multitude of organisations are engaged in what is broadly described as catchment planning, and are more focused on the wider aspects of catchment management, making

links to land use planning, and trade-offs between water use and social and economic benefits of various land use activities. Especially important is the non-statutory Natural Resource Management (NRM) system, designed to allocate funding, from the national as well as state governments, for resource management within priority regions. At Commonwealth level this is implemented through the 'Caring for Country' scheme, and is complemented by NRM planning at state and regional levels. Although this is about land management, boundaries for administering the schemes are catchment boundaries, and there are 14 catchment-based regions in Queensland, although these are grouped somewhat differently from the 22 surface water management areas for water resource planning. Within Queensland, there is a framework for regional NRM,[163] and each region produces its own NRM Plan. In South East Queensland (SEQ), the Plan was produced by the SEQ Regional Coordination Group, but published by the Department;[164] in Condamine, the Condamine Alliance led the process and produced the Plan.[165]

NRM Plans include long-term aspirational targets, medium-term and short-term targets, and strategic management actions. In Condamine, targets are set for water, nature (biodiversity) and land. In South East Queensland, a much more diverse catchment including Brisbane, there are targets for other elements including air quality, and coastal and marine waters. In recognition of the complex planning environment, there are specific targets for bringing together institutional and individual stakeholders, as well as engaging with indigenous groups.

The NRM plans are broader than the WRPs; they are not solely concerned with water, and in relation to water they look at wider matters than allocation. They address water quality and biodiversity and ideally perform an integrative function, and utilise a variety of other organisations that may focus on water; see, for example, in Condamine, the Condamine Catchment Management Association.[166] In SEQ, from a much wider list, see, for example and especially, the SEQ Healthy Waterways Partnership[167] and SEQ Catchments.[168]

2.8.6 Public participation

There is statutory provision for consultation in the QWA, and this section will focus on participation in the statutory processes, whilst noting that the catchment-based NRM activities are an important mechanism for engagement; as would be expected in a

---

[159] The Act provides both for a volumetric limit, which is the total maximum that could be taken in any one year (QWA s.120B), and the nominal volume (QWA Sch.4), which essentially provides for a proportionate share in the entitlement, either for all holders, or, where there is a ROL, within each priority group.

[160] New South Wales–Queensland Border Rivers (Qld) Act 1946 11 Geo.6 No.16 (as amended) Sch.1.

[161] DNRM (2012).   [162] Water Act 2007 s.54.

[163] DNRM (undated).   [164] Queensland Government (2009).

[165] Condamine Alliance (2010).

[166] 'Condamine Catchment Management Association' see http://www. condaminecatchment.com.au/; this was initially set up to take forward earlier planning arrangements for the Murray–Darling.

[167] 'Healthy Waterways Partnership' see http://www.healthywaterways.org/ AboutUs/PartnershipOffice.aspx and Chapter 4.

[168] 'SEQ Catchments' see http://www.seqcatchments.com.au/.

jurisdiction with a small population and extensive rural areas, they may involve the same individuals and groups.

For the WRPs,[169] the Minister must publish a proposal for public consultation; there is a minimum period of 30 days for responses, and the Minister may decide to issue a further public notice. Notices are issued to local governments and to any entities that the Minister considers appropriate; local governments must make such information available. When responses have been received, the Minister should make the draft Plan available, with an overview report. Again, a further draft may be issued. The Minister must consider all submissions, and report on the consultation process – as in Scotland, a useful and important provision that lets stakeholders see how their responses were addressed. The Minister must also produce annual reports. The final WRP is approved by the Governor in Council, when it has legislative status. This process has been reduced in recent years, to make it less burdensome; especially, the requirement to establish a community reference panel.[170] The tension between swift decision-making and full engagement is clearly seen in most jurisdictions, but in Australia, unlike the EU states, there are few external pressures to prefer the latter.

For the ROPs,[171] again, public notices will be issued of the intention to draft an ROP. Owners of infrastructure must be notified in order to provide information. Again, there must be an overview report. ROPs are prepared by the chief executive, but there is now a process for concurrent consultation, and concurrent approval, along with a new or amended WRP. For ROPs, if submissions have been made in relation to changing water allocations, environmental management rules, water sharing rules or an implementation schedule, a referral panel must be established to consider these and report to the chief executive,[172] unless the chief executive intends to amend the ROP in accordance with the submission. Again the Plan is signed off by the Governor in Council. The consultation reports give an indication of the nature of issues raised by stakeholders and whether the Plan was amended in response.[173] It is perhaps unfortunate that these panels no longer have a role in the wider WRP process.

2.9   SOUTH AFRICA

In South Africa, a statutory system for integrated water resource management and catchment planning is provided for through the NWA, which addresses all elements of the freshwater resource.

The NWA itself was drawn up in the context of the post-reform policy framework, setting out basic principles of sustainability, equity, efficiency and participation, and recognising the need to redress historic imbalances of race and gender for access to resources.[174] It is a well-designed legislative framework providing for IWRM at national and catchment scales, through the National Water Resource Strategy (NWRS) and Catchment Management Strategies respectively, along with a proportionate system of integrated water use licences, covering all aspects of resource use through different tiers of control. As noted above, the definition of water resources in the NWA includes surface waters, groundwater and also wetlands, though it does not extend to coastal waters. Overall the NWA has provided a structure for water resource management that enabled a very proactive and water-focused reform agenda to be carried out in a challenging economic and social environment, but one in which there was a high incidence of political will. Currently the NWA works with the Water Services Act (WSA),[175] though there is a proposal from the Department of Water Affairs (DWA), also discussed in the second NWRS (NWRS2), to merge the two Acts into one complete legislative framework, along with certain other legislative and policy reforms.[176] South African experience in IWRM and the associated operational areas of water quantity and quality management provide many excellent examples of the iterative nature of the IWRM process, the way that process can develop over time, and the hurdles and difficulties encountered.

The purpose of the Act is stated as being to 'ensure that the nation's water resources are protected, used, developed, conserved, managed and controlled in ways which take [the following] into account …'.[177] There follows a list of non-exhaustive factors, including the basic human needs of present and future generations, equitable access, redressing past discrimination and 'efficient, sustainable and beneficial use'. These factors are not stated to be in priority order, but the NWA also lists the provision to be made by the NWRS, beginning with the Reserve, then international obligations, and then other water needs.[178] The first NWRS considered that this was an order of priority.[179] The NWRS2 is more focused on the general principle of equitable reallocation, within which basic human needs are a starting point, and de facto prioritised, but only a starting point.

The NWA established the Government as the 'public trustee of the nation's water resources',[180] reformed historic provision for water rights (Chapter 3), and created a whole new series of water

---

[169] QWA Chapter 2 Part 3.

[170] Previously, s.41 QWA; repealed, Water and Other Legislation Amendment (Qld) Act 2011 No.40 s.15.

[171] QWA Chapter 3 Part 4.

[172] Referral Panels can be established by the chief executive, QWA s.1004, and will have three individual members.

[173] See, e.g., DERM (2011).

[174] See especially DWAF (1997). This incorporates 28 Principles for Water Management developed by the post-apartheid regime.

[175] Water Services Act of 1997 (WSA); and see Chapter 5.

[176] National Water Policy Review General Notice No.888 of 2013 (NWPR 2013); DWA (2013) (NWRS2). These proposals will be considered further in this chapter, and in Chapters 3 and 5.

[177] NWA s.2.   [178] NWA s.6.   [179] DWAF (2004) Chapter 2.

[180] NWA s.3.

management institutions, including especially the Catchment Management Agencies (CMAs) and Water Users' Associations (WUAs). The DWA is also a water institution and continues to fulfil water resource management functions. In the longer term, the Act envisaged that the CMAs would be the primary institution for resource management, with a very ambitious agenda to have stakeholder-led planning and allocation processes; this has proved difficult to achieve, and under the NWRS2 significant changes are being made to the structure of the CMAs, though the aspirational policy goal remains.

The NWA requires the establishment of the NWRS, to be prepared by the Minister and reviewed five-yearly,[181] and then Catchment Management Strategies in water management areas. Initially, there were 19 such areas, each with a putative CMA;[182] but only two have been established and the NWRS2 intends that these will be merged into nine. These water management areas, similar to the EU's RBDs, are essentially groups of catchments that maintain (broadly) hydrological boundaries in line with good IWRM practice, but ideally will achieve the economies of scale and expertise to enable them to take on their challenging roles. The Act sets out the content of the strategy: *inter alia* it must provide for the Reserve, international obligations, future needs, and water management areas.[183]

The Reserve is an innovative feature designed to provide on the one hand for human needs (drinking, food preparation, and personal hygiene) and on the other, for ecological needs. The human needs component of the Reserve is calculated on the minimum requirement of 25 litres per person per day (LPD), which is also the Basic Water entitlement;[184] the ecological Reserve is more complex and has not yet been completed in the NWRS2. All water management institutions must give effect to the Reserve when exercising their functions.[185] In the 1997 White Paper, and other subsequent policy, it is stated that only the Reserve, providing for basic human and environmental needs, should be available as of right; the Reserve is protected against other uses, all of which are subject to equitable principles.

The NWA specifically provides for allocations to meet international agreements,[186] and these are then incorporated into the NWRS and the Catchment Management Strategies as a priority water use, second only to the Reserve. South Africa shares four

major river systems with six immediate neighbours, affecting 11 of the 19 original catchments, and has a number of agreements relating to international watercourses, within a framework provided by the Revised Protocol on Shared Watercourses in the Southern African Development Community.[187]

There is also provision in the Act,[188] and supplemented by detailed policy, for resource protection.[189] This will establish Resource Quality Objectives (RQOs) based on different management classes for water. A six-fold ecological classification feeds into three management classes (class A, to minimally used/impacted; class B/C, to moderately used/impacted; and class C/D, to heavily used/impacted; classes E/F are considered unacceptable and should be remediated). The system of ecological classification is not dissimilar to that of the EU WFD, except that there is no presumption that all waters should reach class B; determination of the final ecological Reserve is dependent on this process. The highest management class requires the most protection and hence the most stringent objectives, and the highest level of Reserve flow; see also Chapter 4.

2.9.1 The National Water Resource Strategy

Under the Act, the NWRS must provide for certain matters, including the Reserve, international obligations, and strategic water use; estimates of current and future water requirements and areas of surplus or deficit; principles of conservation and demand management; and holistic and integrated catchment management.[190] The NWRS identifies imbalances in supply and demand, and is a framework for more detailed analysis. It does not determine the amount of water available for allocation by a CMA.

The first NWRS was a comprehensive document, including a preliminary estimation of the Reserve; identification of inter-basin transfers; a reconciliation of supply and demand, nationally and at catchment level; an estimation of likely future needs till 2025; and the potential for demand management and new infrastructure development. It also included an overview of the 'strategic perspectives' developed for each of the original 19 water management areas, to be used until CMAs were established and operating, and a national water balance. In the NWRS2, these have been replaced by reconciliation strategies looking at the supply and demand balance in the nine

---

[181] NWA s.5.
[182] Establishment of the Water Management Areas as a Component of the National Water Resource Strategy General Notice No.1160 of 1999 establishes the 19 water management areas, each of which will have a CMA. NWA ss.8–11 provide for Catchment Management Strategies.
[183] NWA s.6.
[184] The provision for Basic Water, including the Free Basic Water Policy, will be discussed in Chapter 5.
[185] NWA s.18.
[186] NWA s.2, purpose; s.6, contents of the NWRS; ss.102–108, on establishment of international bodies.

[187] The Revised Protocol on Shared Watercourses in the SADC Region (2001) is aligned with the UN Convention on the Law of the Non-Navigational Uses of International Watercourses (1997) and came into effect in 2003. South Africa has signed and ratified the UN Watercourses Convention. SADC was originally formed in 1980 and constituted in its present form in 1992; South Africa acceded in 1994. 'South Africa's interaction with SADC' see http://www.dfa.gov.za/foreign/Multilateral/africa/sadc.htm.
[188] NWA ss.12–18.   [189] DWAF (1999); see also DWAF (2007).
[190] NWA ss.5–7.

new areas; it is intended that the national water balance will be revisited in the next edition.

The NWRS2, as befits a high level strategy, reviews and analyses progress to date and sets out proposals for the next planning period. Although the NWRS should be revised every five years, almost ten years elapsed between the first and second versions. Five years may simply be too short a time for such a major production, given the many pressures on the Department's resources.

Certain themes and policy imperatives are emphasised throughout the NWRS2. One of these is water conservation and demand management, which should be the first priority for all water management institutions, as it is likely to be less costly than most other alternatives, such as new surface water storage schemes, water transfers or desalination. Indeed, an emphasis on conservation and demand management, including wastewater treatment and reuse, should be a priority for all states, in developed and developing countries, in a water-scarce world (and see Chapter 5). The NWRS2 is clear that there are water deficits in many catchments and that new surface water supply is not usually an option, although groundwater resources may still be available. There is a separate strategy on groundwater,[191] which suggests that it is especially appropriate for small-scale and rural domestic use, where options such as desalination (or indeed conventional surface water supply provision) will not be affordable.

2.9.2 Catchment Management Agencies

Strategically, CMAs are the most important of the water management institutions set up under the NWA.[192] The CMAs are bodies corporate, with a general duty to redress discrimination and achieve equitable access. Their initial functions are to investigate and advise on the use of the resource; to develop the strategy; to coordinate activities of persons and institutions; to coordinate the NWA with the implementation of the Water Services Act; and to promote community participation.[193] These functions can then be extended, most importantly, to regulatory functions including issuing water use licences, in the context of reallocating water for greater equity.[194]

The policy context made it clear that this would always be a gradual process;[195] meantime, a more top-down approach might be required. The early guidelines emphasised the need for a clear legislative structure that sets out criteria for institutional

decision-making, rather than leaving wide discretion to Ministers. For example, under the Act CMAs can be established by community initiative and not solely by Ministerial action.[196]

Once a CMA is established, the Minister does have wide powers to intervene in the event of mismanagement or failure to perform;[197] he may issue directives, withhold funds or take over any functions, and has powers of disestablishment. He may not delegate his power to appoint CMAs.[198] The Minister may *disestablish* the CMA entirely, for more effective resource management, because the agency is not operating effectively or 'because there is no longer a need for' the agency.[199] Given its nature and purpose it is hard to see how, once established, a functioning CMA could be redundant. The Minister also has some regulatory powers, for example to set the number of Board members or their remuneration.[200]

Although the policy context is well developed, and includes detailed guidelines on preparing catchment management strategies,[201] only two of the original 19 CMAs have been fully established – the Inkomati and the Breede-Overberg; each of these has published Catchment Management Strategies,[202] although clearly these will need to be revised following the expansion of the catchments. In practice therefore, most activity is taking place through the DWA national and regional offices.

The Catchment Management Strategies must include water allocation plans, taking into account statutory factors including existing lawful users, the need to redress past discrimination, efficient and beneficial use and socio-economic impact.[203] The need to reallocate water equitably in the face of increasing pressures and uncertainty has been problematic, despite provision in the NWA for compulsory water relicensing and a policy framework to support emerging farmers.[204] One of the areas currently proposed for reform is to strengthen the role of equity in reallocation. Furthermore, it is likely that when the ecological Reserve has been implemented, areas that were in balance may be in deficit; therefore a proper assessment of the Reserve taking this into account should be a high priority for a CMA.

2.9.3 Participation

In South Africa, public participation is a general political issue. One of the purposes of all the post-apartheid legislation is to redress historic inequalities, and one of the tools to do so is participation in decision-making. The drafting of the South African Constitution was itself held out as an exercise in

---

[191] DWA (2010).

[192] NWA ss.77–90 and Schs.3 and 4 apply to CMAs; ss.72–76 establish Minister's powers re the CMAs.

[193] NWA s.80.

[194] NWA s.73 empowers the Minister to assign these additional functions to a CMA, by notice and after consultation.

[195] See, e.g., DWAF (undated).

[196] NWA s.78(1).   [197] NWA s.87.   [198] NWA s.67.

[199] NWA s.88.   [200] NWA s.90.   [201] DWAF (2007a).

[202] Inkomati CMA (undated), Breede-Overberg CMA (2010).

[203] NWA s.27. These are also considerations for the issuing of licences and will be addressed further in Chapter 3.

[204] See DWAF (2004a). The reallocation process will be considered in Chapter 3.

participation, and the DWA has produced guidelines to facilitate stakeholder engagement.[205] One of the functions of a CMA is to provide community representation in water management, and the Act provides a detailed procedure for the CMA's governing board, to be appointed by the Minister 'with the object of achieving a balance amongst the interests of water users, potential water users, local and provincial government and environmental interest groups'.[206]

Some analysis was done of the CMA process in the first few years. McConkey *et al.* reviewed the process of setting up the CMA in the Breede catchment,[207] and identified various uncertainties, including the lack of a finalised water pricing policy and finalised figures for resource availability and the Reserve. Pegram and Bofilatos considered in more detail the composition of CMA Governing Boards, looking at the Inkomati CMA.[208] The authors considered that water resource classification and compulsory licensing should stay with DWAF to avoid difficulties of both capacity and stakeholder support, as the compulsory licensing regime will involve unpopular reallocations.[209]

Anderson conducted empirical research amongst 62 disadvantaged stakeholders and considered that they found it difficult to access meetings etc., and there were cultural problems with western modes of information distribution and with meeting styles.[210] There were unrealistic expectations, and participation fatigue. Facilitation and dispute resolution were best carried out by local NGOs and not outside consultancy firms. Perhaps most worryingly, some stakeholders, especially agricultural users, did not see the benefits of the CMA. They did not wish to pay charges for its operation, and were at risk of withdrawing from the process.

One mechanism for participation at local scale is WUAs, which exist in many jurisdictions.[211] In South Africa, WUAs are cooperative associations and do not usually have regulatory or management functions; they can be established by the Minister[212] but are more likely to be proposed by interested parties.[213] They are usually, but not always, associations of irrigators. Existing irrigation boards, subterranean water control boards and stock watering boards were intended to convert to WUAs, but this is still ongoing. The NWRS2 indicates that some 90 WUAs now exist, but another 129 (mainly irrigation) boards remain to be rationalised. The DWA intends to support the establishment of WUAs, in particular to assist emerging farmers to access subsidies.[214]

One further mechanism is Catchment Management Forums, which work with the CMAs at that scale. The NWRS2 suggests that they should not have statutory functions, but rather bring together and represent a wider range of stakeholders, similar to the Area Advisory Forums in Scotland. The DWA will support and fund these until CMAs are running.

2.9.4 Future policy and law reforms

Given its extent, it seems sensible to end this section by outlining the extensive reform programme being proposed by DWA.[215] The intention to merge the two principal Acts has been noted. It is relatively unusual to have one legal instrument covering both water resources and water services, but the reasons given are to enable DWA to better manage the technical aspects of service delivery by local governments, to ensure the proper management of the whole water value chain, and to remove inconsistencies. In future, they propose a single National Water Strategy also covering service delivery. Water pricing and economic regulation should apply through the whole value chain, and it will be clarified that the policy of Free Basic Water (Chapter 5) should only extend to the indigent poor; others should pay for their water, with the aim of being able to increase the amount of basic water supplied. To improve equity in water use and allocation, a principle of 'use it or lose it' will be introduced, for all existing water entitlements. Such a policy is not without its problems, but will enable reallocation of water being held, but not used.[216] In addition, the principle of equitable use will be given a higher legal standing amongst the criteria used in reallocation of water licences, and water trading will be abolished. Finally, the 12 Water Boards, which are responsible for bulk supply but also sometimes function as water services providers, will be transformed into nine regional water utilities, which will manage infrastructure and take over the functions of the Water Trading Entity.[217] This is an ambitious programme of reform to add to the ongoing activities around the Reserve, RQOs and reallocation through compulsory relicensing.

2.10 OTHER WATER MANAGEMENT REGIMES

The other areas of national law directly concerned with the management of water, identified in Chapter 1, were the management of floods and droughts, the management of coastal

---

[205] DWAF (2004b).   [206] NWA s.81(1).

[207] McConkey *et al.* in Ostfeld and Tyson (2005).

[208] Pegram and Bofilatos (2005).

[209] DWAF (2007a), and see Chapter 3.   [210] Anderson (2005).

[211] For a discussion of the problems that can arise in the establishment and operation of WUAs, see Allan (2005).

[212] NWA s.92.   [213] NWA s.91.   [214] NWRS2 Section 8.56.

[215] NWPR 2013.

[216] 'Use it or lose it' is part of the system of prior appropriation found in the western USA, and may encourage inefficient use; see Chapter 3.

[217] The Water Boards will be addressed in Chapter 5; bulk transfers and water trading will be considered in Chapter 3.

waters, and the regulation of dams and river works. None can be fully explored in their own right, but each will be mentioned briefly, as each should be considered in the IWRM process.

2.10.1 Management of coastal waters

The law of the sea generally is outwith the scope of this book, which is necessarily restricted to management of freshwater resources. However, flowing water does not recognise lines on a map. There is an international commitment to sustainable and integrated coastal management through Agenda 21, and all of the jurisdictions studied have extensive coastlines of major economic, social and environmental consequence.

The WFD specifically provides for the management of 'transitional' (estuarine) waters, and coastal waters up to one nautical mile from the low tide mark; in Scotland, as noted, WEWS extends the Directive controls to three nautical miles from shore. There is also a significant body of other statutory and non-statutory provision for coastal management, addressing both land use activities and the use of the marine resource itself. The EU has policy around coastal zone management,[218] has published a marine strategy directive,[219] focused on ecology, and is now proposing a directive on marine spatial planning.[220] This last is indicative of a general trend to try to better manage the competing uses of the marine environment, especially fishing, shipping, and now increasingly renewable energy, whilst still protecting marine ecosystems. In the UK and Scotland there is ICZM policy,[221] implementing legislation for the marine strategy directive,[222] and primary legislation already existing to provide for marine spatial planning.[223] The extension of control of land-based activities into coastal waters under the WFD is beneficial, to improve coordination with freshwater management.

In federal Australia and in Queensland, likewise, there is plenty of law and policy. The QWA does not apply to coastal waters, but the Marine Parks Act 2004[224] and the Coastal Protection and Management Act 1995[225] apply. The latter provides *inter alia* for control of coastal zone activities that do not require planning consent. The former enables designated marine parks, including the Great Barrier Reef, in line with the principles in the Commonwealth Great Barrier Reef and Marine Parks Act

1975.[226] Protection of the Reef is a driver for some environmental controls in Queensland, and NRM plans in coastal areas address coastal issues, but there are significant tensions between mining and farming, and environmental protection (and see Chapter 4).

In South Africa, coastal waters do not fall within the definition of water resources under the NWA, although estuaries do, and water use licences cover discharges to sea through outfall pipes. The Marine Living Resources Act 1998[227] makes provision for marine parks, and more recently, the Integrated Coastal Management Act 2008[228] clarifies public rights in coastal areas, and improves controls over pollution as well as creating new mechanisms for development control. So again there is a full complement of law and policy, meeting international obligations, but with limited connectivity to freshwater regimes.

2.10.2 Floods and drought

Floods and droughts are the two extremes of water management. They bring different problems, but both involve emergency planning, and both are affected by climate change. Drought responses may require reallocation of water and reprioritisation of water uses; Chapter 3 will also be relevant and Chapter 5 will consider emergency provision for water services. This section will therefore focus on flooding, which requires planning processes, and participation mechanisms, before, during and after flood events.[229]

The WFD makes reference to mitigating the effects of flood and drought, but it is not a management tool for these events. There is a separate Floods Directive,[230] which establishes a planning process for risk assessment and management, operating on the same timescale as the WFD and explicitly linked into the RBMP process.

In England, there is an extensive and complex planning and policy environment. Flood defence is the responsibility of the EA,[231] which produces statutory Catchment Flood Management Plans, which will feed into the RBMPs. The EU Floods Directive is implemented by regulations,[232] but in addition the Floods and Water Act 2010 brought in new processes for flood and coastal erosion following very severe floods in 2007.[233] Local authorities manage developments and control building on flood plains;

---

[218] Recommendation 2002/413/EC.   [219] Directive 2008/56/EC.

[220] European Commission (2013).

[221] Scottish Executive (2004), DEFRA (2008).

[222] Marine Strategy Regulations SI 2010/1627.

[223] Marine and Coastal Access Act 2009 c.23; Marine (Scotland) Act 2010 asp.5.

[224] Marine Parks (Qld) Act 2004 No.31.

[225] Coastal Protection and Management (Qld) Act 1995 No.41; and for the policy context, see 'Coastal Management in Queensland' http://www.ehp.qld.gov.au/coastal/management/.

[226] Great Barrier Reef and Marine Parks (Cwlth) Act 1975 No.85; and for the policy context, see 'Australian Government Marine Protection' http://www.environment.gov.au/coasts/index.html.

[227] Marine Living Resources Act 1998 No.18.

[228] Integrated Coastal Management Act 2008 No.24.

[229] See, on links to IWRM, UNESCO (2009), UNECE (2000); and on legal issues, Associated Programme on Flood Management (2006).

[230] Directive 2007/60/EC.   [231] WRA s.2.

[232] Flood Risk Regulations SI 2009/3042.

[233] Floods and Water Management Act 2010 c.29. See also Pitt (2008) and DEFRA/EA (2011).

they have new responsibilities now under both the Floods Directive and domestic law. Surface water flooding is an increasing problem throughout the UK, as in other countries, and will be considered in Chapter 5.

In Scotland, flood defence is primarily a local authority function. The Floods Directive, as with the WFD, was implemented in primary legislation,[234] and made some further reforms of the domestic law, including a duty on SEPA to assess the possibility of using natural features (such as wetlands) to manage upstream flood events and reduce the pressure downstream. In general, the situation in Scotland is better than in England, with less population pressure and less evidence of planning permission being granted contrary to SEPA's advice. In England, flood defence rests with the EA and this is considered to be part of the problem; local authorities grant planning permission, but are not responsible for the works to protect homes from flooding.

In South Africa, the NWA empowers water management institutions to make information available regarding floods and droughts. There is a requirement for plans showing the 1 in 100 year floodline before the establishment of a township. Catchment Management Agencies have powers to temporarily control water use in periods of shortage.[235] Generally, both flood and drought are part of disaster management. The Disaster Management Act[236] provides for frameworks operating at national, provincial and municipal level, and a National Disaster Management Centre, under the control of the Department of Cooperative Governance. As in Queensland, fire is another source of disaster requiring emergency planning.

In both the Commonwealth and Queensland, recent water reforms, including the Water Act 2007, have been prompted by severe drought; these affect irrigation and urban water services (Chapters 3 and 5). In Queensland, as in South Africa, there is a Disaster Management Act,[237] which establishes emergency powers at state and local level and specifically addresses floods as well as other natural hazards. Queensland suffered very severe flooding in 2010/11; as in England, it led to an inquiry and report.[238] As in England, the recommendations include better planning, modelling and access to information. It has also led to threatened litigation around the operation of a dam, discussed in the next section. All the recommendations in the report have been enacted following commitment made by the Queensland Government; this will lead to new systems for assessing and approving levees.[239]

2.10.3 Dams and river engineering

Dams are a major area of study in their own right, and again cannot be dealt with comprehensively here. They are linked to flow management, ecology, allocation regimes, water services and energy production. In each jurisdiction studied here the management of dams is fully or partially incorporated into the water abstraction licensing, but large dams raise extended social and environmental issues. In the late 1990s and the early part of this century, large dams were perhaps out of favour,[240] but that trend is shifting again.[241]

Dams and other significant river works are likely to require development consent and also environmental assessment; the relationship between these controls and water management will be briefly addressed in the next section. At a strategic level, the IWRM process could provide mechanisms to consider the wider social and environmental matters for prospective dams. In the EU, existing and future 'artificial and heavily modified water bodies' have special provision under the WFD. Whilst the environmental and social impacts are likely to be managed through other strategic regimes, water management rules are likely to provide for dam safety. This raises many legal issues; in Queensland, at the time of writing, there is the possibility of the largest class action in Australian legal history over the management of the Wivenhoe Dam in the 2010/11 floods, with a suggestion that the operating engineers breached their duty of care. The action, if it proceeds, will be brought against the engineers and also the state government and SEQ Water, which owned the dam.[242] The potential for harm and the issues around liability are compounded if the dam holds 'dirty water' (for example, mine tailings).

In Scotland, impoundments and river engineering works are both controlled activities under the CAR and will be covered by integrated water use licences. Large dams for hydropower schemes are authorised under the Electricity Act.[243] The Reservoirs Act 1975[244] applied to large dams in Scotland and in England, but is being replaced in Scotland.[245] In England, the EA is the responsible authority for authorising impoundments,[246] and now for reservoir safety.[247] In both jurisdictions planning permission and environmental assessment are required for dams and engineering works, and in both there is a recent tendency to remove decisions over major infrastructure projects from local control. There is also an emerging focus on small-scale hydro,

---

[234] Flood Risk Management (Scotland) Act 2009 asp.6.
[235] NWA Sch.4.   [236] Disaster Management Act 2002 Act No.57.
[237] Disaster Management (Qld) Act 2003 No.93.
[238] Queensland Floods Commission of Inquiry (2012).
[239] Land, Water and Other Legislation Amendment (Qld) Act 2013 No.23; and see the Explanatory Notes to the Bill, p.1.

[240] See, for a critical and very influential analysis, World Commission on Dams (2000).
[241] World Bank (2004). World Bank (2010), suggesting the Bank will continue to fund high risk infrastructure, especially hydro power.
[242] 'Wivenhoe Dam Class Action' see https://www.imf.com.au/wivenhoe/.
[243] Electricity Act 1989 c.29.   [244] Reservoirs Act 1975 c.23.
[245] Reservoirs (Scotland) Act 2011 asp.9.   [246] WRA Part II Chapter II.
[247] WA2003 ss.74–80.

for self-generation and feed-in to the grid, to meet the climate change agenda.

The Murray–Darling is the most heavily regulated river system in the world and all of Queensland's rivers have infrastructure regulating the flow. Dams and their flow regulatory functions are provided for under the ROPs, and where water is being supplied through infrastructure, ROLs are required (Chapter 3). There are various obligations on the operator of the dam, including safety rules for large dams;[248] the QWA also covers river works.[249] Dams will need consent under the planning regime; the level of Ministerial discretion in Queensland is wide, and there is a tendency (also seen in the UK) to try to remove the largest projects from local control.[250]

In South Africa, as in Queensland, there are already major dams and storage facilities, and little capacity for new major infrastructure.[251] As in Scotland, all water uses are covered by the same regime and the NWA includes diversion, storage and physical alterations to rivers in the definition of 'water use' for which a licence is required.[252] Dams, like other major water projects, require environmental assessment. There is specific provision in the NWA for the management and use of government waterworks,[253] and for dam safety.[254]

2.11 OTHER LEGAL REGIMES AND ENVIRONMENTAL ASSESSMENT

Chapter 1 identified a set of strategic regulatory regimes that support the operation of the water law meta-regime. These include environmental assessment (at strategic and project level), land use planning, nature conservation and biodiversity, pollution control and public participation. Each of these strategic regimes will have a substantial body of associated law and policy, often reflecting international or regional commitments. They may involve plans or strategies that will overlap with water planning in various ways. Public participation, specifically engagement with the IWRM process, has already been examined, and the environmental protection/pollution control meta-regime is separately addressed in Chapter 4. It is not possible to look at most of these areas of law, but this section will briefly touch on environmental assessment, which is essentially a set of procedural requirements for decision-making, rather than a substantive regime in its own right, and affects both IWRM and many operational aspects of water management.

Interesting questions arise from a study of these supporting frameworks, and the mechanisms to integrate them with water resource management planning. These questions arise at two levels: the integration between the planning processes, and the inter-agency (or inter-departmental) liaison in granting consents for water projects.

With reference to the integration of the plans, every jurisdiction requires the water resource plans to be taken account of in other functions, but none of them make any provision as to the primacy of particular plans, and government guidance invariably stresses the need for conflicts to be amicably resolved and, where necessary, to be brought to the early attention of Ministers. Clear specification might avoid shifting conflicts 'downstream', where they may impact negatively on wider stakeholder engagement.

Strategic Environmental Assessment (SEA) is developing in all the jurisdictions, and requires the assessment of plans and strategies such as IWRM is likely to produce. Although it developed later than project-based environmental assessment, it should be considered first as it applies at an earlier stage. It should allow better consideration of all the relevant options, and it is a vehicle for stakeholder engagement. The EU makes detailed and mandatory provision,[255] which risks increasing the administrative burden unless the subject matter of the SEA is carefully circumscribed.[256] In South Africa, SEA is provided for in the National Environmental Management Act 1998 (NEMA)[257] but is not mandatory, which may enable a different, lighter touch but leaves much to the discretion of the decision-maker. In Australia, federal law provides for SEA;[258] if such an assessment is carried out there is a power for the Ministers to then require a less onerous environmental impact assessment at project stage.[259]

At the level of project consent, the primary mechanism in many jurisdictions is the land use planning system, which is another meta-regime that should be linked to, and not conflict with or duplicate, IWRM. The substance of land use planning is beyond the scope of this book, but EIA also operates at project level and is a mechanism for participation. It may be linked to land use planning, or operate separately.

As with SEA, EIA is mandatory for the EU states[260] and the process, and affected projects, are closely prescribed; the UK regimes cannot select the desired form of assessment or set the thresholds.[261] Major water projects require EIA. In Queensland,

---

248 Water Supply (Safety and Reliability) (Qld) Act 2008 No.34 Chapter 4.
249 QWA ss.266–269.
250 State Development and Public Works (Qld) Act 1971 No.55.
251 NWRS2 Chapter 4.   252 NWA s.21.   253 NWA ss.109–116.
254 NWA ss.117–123.

255 Directive 2001/42/EC; Environmental Assessment of Plans and Programmes Regulations SI 2004/1663; Environmental Assessment (Scotland) Act 2005 asp.15.
256 See Carter and Howe (2006).
257 National Environment Management Act 1998 No.107 (NEMA); and see DEAT (undated).
258 Environmental Protection and Biodiversity Conservation (Cwlth) Act 1999 No.91 (EPBC Act).
259 EPBC Act s.146.   260 Directive 2011/92/EU.
261 Town and Country Planning (Environmental Impact Assessment) Regulations SI 2011/1824; Town and Country Planning (Environmental Impact Assessment) (Scotland) Regulations SSI 2011/139.

by comparison, the environmental assessment rules are relatively flexible, but there is some prescription around the process.[262] It can also be noted that in Queensland the QWA makes detailed provision as to the coordination with land use planning legislation, where the chief executive under the QWA may be either an assessment manager (first decision-maker) or a referral or concurrence agency (consultee) in works that require consent under both sets of rules.[263] It is unusual, and helpful, to have such specific provision in the water legislation. In South Africa there are rules under NEMA,[264] which as in the UK specify thresholds for projects and give more detail on processes. Major water infrastructure projects should certainly be assessed, and it is desirable to specify affected projects along with any thresholds of scale in the legislation. These matters should not be left to Ministerial discretion.

2.12 CONCLUSIONS

What conclusions can be drawn from the above analysis? This chapter set out to examine the global policy context of IWRM and then assess the provision made in the four comparators. Each of the jurisdictions has a system for management of water resources, based on catchments and agglomerations of catchments. In every case, there is a holistic approach to the resource, insofar as groundwater and surface waters are managed together. Every jurisdiction has the basic elements of IWRM.

States may implement IWRM by a combination of primary and secondary legislation and guidance. The use of primary legislation, at least as a framework, allows specification of high level duties and/or principles, which can then apply not just to IWRM but to the operational regimes within its strategic framework, especially water allocation and pollution control. Primary legislation also ensures adequate debate by legislatures and indicates the importance of the underpinning policy. Such legislation can then contain definitions of the water environment, again applicable to the operational regimes. Groundwater should be included in this, as may be wetlands, diffuse surface waters and coastal waters.

Each jurisdiction produces output from the management process, which may be plans or strategies. These make an assessment of the resource base and then set conditions and targets. In each case, the planning process is intended to be inclusive and participative for user groups and the wider community; in each case, the plan – at least – sets the context for future regulatory decisions in the core areas of water allocation and water pollution.

In South Africa and Queensland, regulatory powers rest with government departments. In the UK, there are separate agencies. This distinction affects the structure of regulatory regimes. In the UK, the regulators for allocation and pollution control are also the lead authorities for IWRM, and the water services providers also have an important planning role. In South Africa, water resource management is intended to be devolved to the stakeholders under a statutory process. In Queensland, there are different planning processes: allocation is statutory and led by the department, and resource management is led by user groups.

The regulatory role of the plans is variable. In Queensland, the WRP/ROP process is directly regulatory; the finished plans have the status of secondary legislation and create a finalised allocation regime. The NRM system by contrast is primarily managerial. It sets targets, but enforcement is achieved primarily by financial, rather than command and control, measures. In South Africa, the CMS will also set out principles for allocation and, in addition, regulatory powers are intended to be progressively devolved. In the UK, the RBMPs are not directly binding, but the programmes of measures to achieve them will be implemented through revised licensing conditions, and these in turn are operated by the authority leading the planning process.

Proper planning processes do not eliminate conflicts between water users; however, IWRM provides information on the resource base and the demands upon it, enabling greater transparency and thus increasing the possibility of adequate resolution. It is desirable to have explicit provision for identification of and engagement with stakeholders. In every jurisdiction there is separate legislative provision for environmental assessment and for access to information, both of which are pertinent to water management and assist with participation.

On explicit provision for links to other planning processes, in South Africa, NEMA does address this at strategic level, and in Queensland, the Sustainable Planning Act provides detail at project level. More broadly, IWRM can and should provide an integrative framework, a forum for bringing stakeholders together. Every jurisdiction seeks to do this. Ideally, the legislative framework for IWRM should identify all linked agencies and statutory processes. This would be preferable to making linkages through more complex policy statements, and provide

---

[262] Major projects will have an Environmental Impact Statement under the State Development and Public Works Organisation (Qld) Act 1971; EIS is also provided for under the Sustainable Planning Regulation 2009 SL No.280 Part 6, and if an EIS is carried out under that Act or federal law then the 1971 Act may be disapplied.

[263] QWA ss.966–972, Sustainable Planning (Qld) Act 2009 No.36.

[264] Regulations in Terms of the National Environmental Management Act Chapter 5 2006 No.385, No.386, No.387. Reg.385 is the principal regulation setting out procedures, whilst Reg.386 and Reg.387 are the lists of activities controlled and competent authorities.

Table 2.1 *Key findings Chapter 2: IWRM and river basin management*

	South Africa	Queensland	England	Scotland
Regulator	Department	Department	Agency	Agency
Purpose/ principles	Human needs; equity; redress discrimination; efficient, sustainable and beneficial use.	Sustainable management; efficient use; ecologically sustainable development.	[IWRM in secondary law, no high level provisions. Conservation/resource management duties in Environment Act.]	Protection of the water environment; prevent deterioration; beneficial, sustainable and efficient use.
Integrated water management	Catchment based. Surface and groundwater; estuaries; wetlands.	Catchment based. Surface and groundwater; overland flow.	Catchment based. Surface and groundwater; inland and coastal water.	Catchment based. Surface and groundwater; inland and coastal water; wetlands.
Single planning structure?	Yes National/catchment	No <u>Allocation</u> Quality/resource management	Yes Regional/sub-catchment	Yes National/catchment
Executive function	Department ↓ Stakeholders (progressive)	<u>Department</u> Stakeholders	Agency	Agency
Status of plan	Regulatory Indirect (progressive) Sets targets Licence	Regulatory Direct <u>(Plan as Licence).</u> Managerial Sets targets £ Incentives.	Regulatory Indirect Sets targets	Regulatory indirect Sets targets

welcome clarity. There is no clear statement in any jurisdiction as to the status of other plans in relation to IWRM. Clear direction from legislators would be indicative of a more systematic approach, not just in water management, but generally in environmental regulation.

A state seeking to reform its water law will not always choose to begin with resource management frameworks. It may be desirable to begin with reform of an operational area, and this will be considered again in Chapters 3 and 6. Nonetheless, water resource management is a global policy goal and may be a national priority; it may be introduced with no, or only limited, concomitant operational reforms. Particularly if IWRM is being implemented on top of existing structures, it has potential to play a role in both integrating and rationalising water management functions. Many relevant plans and processes related to IWRM have water as a focus, such as management of dams, floods, or coastal waters. If these could be integrated it would rationalise the number of separate plans and provide a welcome focus for disparate processes. This is certainly the case in the UK, where many of the related processes are already operated by SEPA and the EA and could be produced as subject-specific 'sub-basin' plans. It has also been identified in Queensland, where despite their non-statutory

nature the NRM plans could provide an integrative focus, being much wider than the WRPs.

When states are establishing IWRM, the primary requirements are to manage the whole resource, on a catchment basis. There should be a clear policy context, with policy goals that might include equity, ecological sustainability, efficiency and integration. It is well established that improved management of and access to water improves health, increases food security and reduces poverty, as well as providing for environmental protection and ecosystem functioning. Broad social objectives therefore provide part of the rationale for IWRM, but narrower environmental goals can also be met or moved towards. States will then need to consider structural and institutional provision, including whether to establish new agencies or bodies. Models for the planning process are, broadly, that the process is led by a government department or agency, or led by user groups and regulated by a government department or agency. Initial functions should include assessment of the resource, identification of users and pressures, setting targets or objectives, and establishing participative processes and relationships with users and other stakeholders, including other agencies. As the process develops, if it is to be led by users these functions can be transferred first, followed if that is the intention by regulatory matters such as issuing licences, setting standards, and collecting fees and charges. A staged approach is essential in a process that is fundamentally iterative and developmental, especially where resources are scarce. A well-thought-out system of catchment planning will set a framework for allocation and pollution control, establish monitoring, assist with managing distribution and water services, collect and publish data, and enable information flows and good stakeholder relations. The regulatory role, the degree of integration, and the strength of participative mechanisms are the features that will determine the success of IWRM.

3 Water rights and allocation

3.1 INTRODUCTION

This chapter will address water rights – the legal provision for allocation of water. This is the core of any national water law, the primary functional element following from the strategic planning models discussed in Chapter 2. In turn, the availability of water for instream use is one determinant of water quality (Chapter 4); and the availability of water for consumptive use is a prerequisite for the delivery of water services (Chapter 5). Although water rights can be reformed without bringing in IWRM, the type of data that IWRM provides will be necessary for rational allocation; the institutions and stakeholders are likely to be the same, or at least with significant overlap; and if IWRM has been introduced, some fundamental questions may already have been addressed.

The reform of abstraction rights is also central to the global policy context canvassed in Chapter 1. The multiple stresses on the water environment caused by over-use, rising populations with increasing water services requirements, urbanisation, increasing demand from industry, and increasing degradation of the resource with resultant ecosystems damage, all point to the need to allocate water rationally. This necessitates the reform of pre-existing rights; it is likely that this will require primary legislation, an Act or Code, or at least that primary legislation will provide a more thorough debate and promote consensus.[1]

The global policy frameworks discussed in Chapter 2 have promoted reform of water rights along with the introduction of IWRM, and there is a significant body of recent comparative literature.[2] There will be a pre-existing system for allocation, which in developing countries may be based on customary law. Some writers argue for retaining these customary rules, at least within a pluralist system. They argue that the strengths of customary systems (such as predictability and social acceptability) are not properly recognised in modern laws, and also that reform may disadvantage existing holders of water, especially where they do not have tenure to land.[3] Alternatively, customary systems may perpetuate entrenched inequalities, for example based on gender or ethnicity, or may be unable to adapt to changing pressures on the resource. Others consider therefore that the best approach is to introduce a modern permitting system, which should still be designed to take account of existing customary rights.[4] In countries where prior rights may have the status of property rights and be entrenched constitutionally, this may also be a barrier to reform. In Europe, the European Commission has suggested that, in order to achieve the goals of the WFD, states might be required to reform their systems of water rights.[5]

Salman and Bradlow have concluded that it is generally accepted that a modern water law will allocate water by a permitting system;[6] the same presumption is made by Hodgson for the Food and Agriculture Organization of the United Nations (FAO).[7] All the jurisdictions studied here have such a permitting system, and these are the principal subject of this chapter. Hodgson argues that where resources are scarce it may be better to focus on allocation rather than the broad and complex area of IWRM.

Salman and Bradlow also consider that a modern water law should adopt some form of public trust approach to the ownership of the resource, recognising the state as custodian in the public interest.[8] This is one of the preliminary design questions that may be addressed in a 'water law', whether that law is bringing in IWRM or restricted to water rights. In South Africa and Queensland, there is explicit provision for trusteeship[9] or state ownership;[10] this is not the case in the UK jurisdictions, but such may increasingly be implied by the very establishment of the licensing regime, and by prior legal conceptualisation of the public nature of rights in the resource.[11]

States may establish a hierarchy or prioritisation of water uses in their legislation, or in policy; domestic use is almost always prioritised, and priorities may apply at all times, or only in

---

[1] Hodgson (2006) Chapter 5.
[2] See, e.g., Brun et al. (Eds.) (2005), Hodgson (2006), Salman and Bradlow (2006), Dellapenna and Gupta (Eds.) (2008).
[3] See, e.g., Van Koppen et al. (Eds.) (2007).

[4] See, e.g., Burchi (2005).
[5] European Commission (2012) para.8.10.4.
[6] Salman and Bradlow (2006) Sections 4.2, 4.4.   [7] Hodgson (2006).
[8] Salman and Bradlow (2006) Section 4.2.
[9] NWA s.3 establishes a public trust.
[10] QWA s.19 vests water in the state.   [11] Hendry (2013).

shortage or other emergency. In terms of developing a new framework for allocation that is transparent and participative, and especially where resources are scarce and all demands cannot be met, a clear statement of priorities is likely to be helpful. Increasingly there is a recognised need to allocate water for the environment.

States also wish to set out the principles on which allocation decisions will be based, such as equity, efficiency or sustainability;[12] these may be stated in an overarching IWRM framework. In South Africa, key principles are equity and efficient, sustainable and beneficial use, and as seen in Chapter 2, the intention is to give equity a higher status in reallocation decisions. In Queensland, there are overarching purposes of sustainable management and efficient use, and a high level principle of ecologically sustainable development, which may be used as a test of official actions. In Scotland, the purpose is protecting the water environment, and there is an aim of sustainable, balanced and equitable use. In England, there are conservation and water resource management duties applying to the regulator under the Environment Act.

This chapter begins with a brief discussion of different approaches to water rights, including the riparian principles that existed in Scotland and England, and were exported to both South Africa and Australia. In all the jurisdictions, they have now been overlaid by statute, which will form the basis of the analysis, including the essentials of a generic licensing regime. In Scotland and South Africa, the same licensing regime applies to all uses of water including its abstraction. In England and Queensland, different regimes apply to abstraction and to discharges, but nonetheless there are certain common features, which will also be relevant to Chapter 4. The chapter will also consider the supply of bulk water, and the development of water trading.

3.2   WATER RIGHTS

This book uses the term 'water rights' in a broad sense, to include a variety of rights to access and use water, including those created by common law, and by administrative licensing regimes. The term is not restricted to a secure, severable and tradable property right. Each regime will use different terminologies, used here where appropriate, e.g., both South Africa and Queensland use 'water entitlement' as a general term. Similarly, each regime will use different terms for licences or permits, but again, these are used here as general terms.

Extensive work has been done comparing approaches to water rights and allocation in different legal systems, particularly by Caponera.[13] He identified five great legal systems, each with its own approach to water management, being Chinese, Islamic, Hindu, Soviet and Roman; the Roman and Islamic schools, for different reasons, have had influence across the globe. There are some common themes; water resources tend to be public goods, managed by some public authority (which may be an emanation of the state or of the community). It is a good to which some may acquire exclusive rights, but usually rights of use rather than ownership, and subject to some wider controls in the public interest. Often there is differentiation between flowing (usually surface) water, and groundwater, or water in springs or diffuse waters; the latter are most susceptible to exclusivity. Historical systems that give exclusive and tradable water rights are unusual; these are found in the prior appropriation system in the American west,[14] but make it especially difficult to reallocate water. None of the jurisdictions here have such a system; although market reforms in Australia do create severable and tradable rights, they do so within a planned system.

Within the Roman system of water law comes the subdivision into civilian and common law systems, and within the latter the development of the riparian doctrine. This chapter will briefly examine private rights under riparian systems focusing on Scotland (a mixed system with civilian roots overlaid by common law)[15] and England.[16] In South Africa (also a mixed system)[17] and in Queensland (where the English common law was received in the nineteenth century),[18] there are few vestiges of the doctrine.

3.2.1   Public and private waters

In Roman law, flowing water was *res communes* (common to all, and owned by no-one), but certain rivers were *res publicae* and belonged to the public, or the state. All persons could use the water, and all had access to the banks for this purpose.[19] In Roman law, public rivers were rivers with perpetual flow, whilst torrents (which did not flow all year) could be classified as private waters.

As later jurisdictions developed their law, the tests for a public river changed. Public rivers under the French Civil Code[20] are 'navigable or floatable' and this test is widely used in many civilian jurisdictions. These public waters are part of the public domain and

---

[12] For a comparison of such underpinning principles across the 16 countries studied, see Salman and Bradlow (2006) Section 3.2. Note that they would consider IWRM itself to be an underpinning principle of a modern water law.

[13] Caponera (1992).
[14] For a comprehensive analysis of US water law see Tarlock *et al.* (2009).
[15] See for an historical analysis, Ferguson (1907), or Whitty 'Water Law Regimes' in Reid and Zimmerman (2000).
[16] Getzler (2005).
[17] Rabie and Day 'Rivers' in Fuggle and Rabie (1992).   [18] Fisher (2000).
[19] Institutes of Justinian, Book II Title I, Translated by Lee (1956).
[20] French Civil Code Art.538.

their use will be regulated by permit.[21] Smaller streams, groundwater and springs may be classified as private waters, and owned by the owner of the land. In South Africa, the Roman law test of permanent flow was retained. Navigation was not the primary test, and most rivers were public. Between the mid nineteenth century and the late twentieth century, decisions of the courts and legislatures moved between riparianism, giving more rights to landowners, and civilian principle, giving more rights to the public;[22] but the modern statutory regime has clarified the position.

In England, only tidal rivers are public rivers; the *alveus* is held by the Crown[23] and the public have only limited rights of navigation and of fishing. In Scotland also, the Crown holds a property right in the *alveus* of tidal rivers,[24] as it does in the foreshore and the seabed,[25] and the public have rights of navigation and fishing. The property may be alienated at least in the modern law,[26] but the public right will be protected. In Scotland, however, all navigable rivers are public rivers even where the *alveus* and banks are owned privately.[27] The right to navigate is protected against the landowner[28] but at common law there is no supporting right of access.[29] This issue of access is critical in some circumstances, and water codes should make provision for this. In South Africa, there are statutory servitudes of aquaduct (for drawing water), abutment (for works) and submersion (where this is necessary), and these can be created by agreement, or ordered by the Court, with compensation if necessary.[30]

3.2.2 Riparian systems

The essentials of the riparian doctrine are well known.[31] Landowners, and sometimes their tenants, have certain rights to use the water, subject to the correlative rights of all the other riparians on that watercourse. The only right in the water is a right of use. Once abstracted, the water becomes a corporeal moveable capable of ownership; but if abstracted or diverted for a non-consumptive use and subsequently returned to the stream, it reverts to its previous nature.[32]

The basic principle is that each owner is entitled to the water that flows through their land, undiminished in quantity and quality – the natural flow theory – and to abstract water and use it, for primary purposes. Primary purposes will include, most importantly, drinking water for humans and also for stock; water for washing and other domestic purposes; but not for spray irrigation, or for manufacturing or other industrial use. Where water is extracted for consumptive use there will be some diminution of the quantity passing downstream, so the doctrine incorporates an inherent conflict; but in water-rich regions domestic uses are unlikely to substantially prejudice a lower proprietor, or at least to do so with such frequency that the rule becomes unworkable and requires change. Similarly, return of household wastewater will lead to some impairment of water quality, but as long as this is minor, and does not infringe the lower heritor's right to use for his primary purposes, there is no cause for action.[33] The logical consequence of the doctrine is that an upper proprietor may use all the water for primary purposes, and the lower heritor must accept that result; in its pure form the rule can only stand where there is in general a sufficiency of water. Furthermore, the rule is inadequate to deal with manufacturing, and other industrial uses; even non-consumptive use for power generation will alter the flow and affect downstream and opposite proprietors. The strict doctrine of natural flow has many limitations in logic and practice.

We can contrast the further development of the doctrine in the riparian states in the USA, where reasonableness became the key factor in determining relative rights of riparians. A 'reasonable use' approach also permits the possibility of use of water by non-riparians, and/or on non-riparian land.[34] More generally, it is now recognised that riparianism is not an appropriate doctrine to enable a coherent and prioritised distribution of a precious resource.

In Queensland, like other Australian states, the riparian system was abolished when it became apparent that it was not appropriate for such a water-scarce country; water is vested in the state.[35] Although owners of land adjacent to watercourses still have a preferential position when applying for water licences, and can

---

[21] E.g., in France, the Law of 16 December 1964, cited in Caponera (1992) p.77.

[22] Rabie and Day (1992).

[23] Though the Crown may alienate the right, but the public rights are still protected *vis à vis* the new owner.

[24] *Crown Estates Cssnrs v Fairlie Yacht Slip Ltd* 1979 SC 156.

[25] See *Shetland Salmon Farmers Association v Crown Estates Cssnrs* 1991 SLT 166 for a full debate of the nature of the Crown's right in the seabed.

[26] *Shetland Salmon Farmers Association v Crown Estates Cssnrs* 1991; exactly a hundred years earlier, in *Lord Advocate v Clyde Navigation Trustees* (1891) 19R 174, the Court was less certain of the right to alienate but equally sure that there was a property right.

[27] *Colquhoun's Trustees v Orr Ewing* (1877) 4R HL 116.

[28] *Colquhoun's Trustees v Orr Ewing* (1877).

[29] *Wills Trustees v Cairngorm Canoeing Club* 1976 SLT 162, where the House of Lords clarified that, as in South Africa, one could sail, but not land, down a stretch of navigable river where the bed and banks were in private ownership. See also Land Reform (Scotland) Act 2003 asp.2, giving rights of access for recreational, educational and some commercial activities, including access to and over water, as long as the right is exercised responsibly.

[30] NWA ss.126–136.

[31] For a thorough analysis of the emergence of the doctrine in the USA see Wiel (1919); and in England, Scott and Coustalin (1995).

[32] For a detailed discussion of principles of ownership and ownership rights in water in the context of a system based on civilian concepts, see Hu (2006).

[33] For a detailed analysis of the 'nuisance cases' and the riparian doctrine as regards water pollution in Scotland, see Hendry 'Water Resource Management and Water Pollution' in McManus (2007).

[34] *Pyle v Gilbert* (1980) 245 Ga. 403, 265 S.E.2d 584.

[35] Now, QWA s.19.

abstract water without permission for stock and domestic use, Fisher finds that the modern statutory rules leave little room for the application of riparian principle.[36] Tan agrees that the 'vested in' formula only replaces common law rights to the extent of the statutory controls, but the latter are extensive.[37]

Similarly, in South Africa, the modern law establishes that the Government is the 'public trustee of the nation's water resources' and has a duty to ensure the management of that resource. Historic concepts of private (and discriminatory) ownership of water resources were simply replaced with the introduction of a modern system of resource allocation and control, reverting to the language of the civilian system that pertained before the 1860s. In South Africa too, some riparian rights (but not the name) are preserved along with other, usually small-scale or domestic users' rights, and again these will be addressed below in the context of the modern law.

Neither England nor Scotland makes any positive provision about the state. This has not been an impediment to the introduction of licensing regimes, and may have avoided potential political controversy. However, we can note Tan's view that a clear statement of public property in water, probably expressed as some form of public trusteeship, has certain advantages. It provides clarity for users of the resource, but also, she suggests, it may improve the care with which public servants exercise their management role.[38]

3.2.3   Groundwater and percolating water

In many countries, the historic principle for both groundwater and diffuse or percolating water is that these are owned by the owner of the land on which they are found, and different rules apply than those for waters in channels (whether above or below ground). The historic right of landowners to abstract and use these waters is found in many jurisdictions and indeed legal systems. The Islamic system of water management permits private ownership of wells, recognising the labour involved in digging these.[39] In Spain, groundwater was brought into the public domain in 1985;[40] Caponera, in his comparative work, states that groundwater is the property of the landowner under the French Civil Code;[41] for the English common law, there is an exclusive right to abstract and use, which he cites as stemming directly from Roman law.[42] Clark surveyed the authorities and concluded that in Scotland this is not a right of property as such, but a possessory right of use and control;[43] nonetheless, that right was widely accepted as going beyond the riparian rights over surface waters.

Groundwater and springs are especially significant given their suitability for drinking water, their susceptibility to pollution and the difficulty of remediation. Such waters should be protected from direct and indirect discharge, and from abstraction at a rate faster than recharge. The hydrological cycle is interdependent, and over-abstraction of groundwater affects surface water; the historic law often developed in the absence of an understanding of the hydrological cycle. Over-abstraction may also affect the stability of the ground, and cause saline intrusion from coastal waters. As discussed in Chapter 2, any modern system of water management should include underground water within a definition of those waters that should be controlled; diffuse waters may also be included for a comprehensive approach, or at least a power to bring these into control.

In England and in Scotland, there is now integrated management of groundwater and surface water under the WFD. In South Africa and in Queensland, both semi-arid climates with a high level of pressure on water resources, the law provides for controls over diffuse surface waters and groundwater, recognising their importance as part of the resource available to users and their role in contributing to the flow of surface waters. Where there are prior private rights, whether over surface water or groundwater, there are potential difficulties with deprivation of those property rights.

3.2.4   Prior rights and deprivation of property

If a state introduces controls that limit or reduce property rights, this may give rise to claims for compensation. Often, the right not to be arbitrarily deprived of one's property by the state will be a constitutional issue.[44] It is very common therefore to grant 'grandfather rights' to existing users, to prevent legal challenge; the extent of these rights will have implications for the effectiveness of a new licensing regime.

In the UK, where there is no written constitution as such, the matter will now be covered by the Human Rights Act 1998.[45] Such deprivation of property should be distinguished from

---

[36] Fisher (2000) pp.90–131.   [37] Tan (2002).

[38] Tan 'A Property Framework for Water Markets' in Bennett, J. (2005).

[39] See, e.g., Naff 'Islamic Law and the Politics of Water' in Dellapenna and Gupta (2008).

[40] See, e.g., Irujo Embid 'The Foundations and Principles of Modern Water Law' in Garrido and Ramon Llamas (2009).

[41] French Civil Code Art.552; Caponera (1992) para.5.1. He notes that in some civilian jurisdictions, the relevant Code prevents abstraction or diversion of springs or percolating waters to the detriment of one's neighbours.

[42] Caponera (1992) paras.6.1, 6.2.

[43] Clark (2006); and for a fuller analysis of the property question, Clark (2002).

[44] See, for some comparative case studies discussing the potential implications of introducing abstraction controls, with particular relevance to South Africa, FAO (1999). In the UK there is no written constitution though the Human Rights Act 1998 c.42 now provides special protection for some rights, including property and possessions.

[45] Human Rights Act 1998, incorporating the European Convention for the Protection of Human Rights and Fundamental Freedoms (1950) (ECHR).

arguments around the 'human right to water' (and indeed sanitation), which will be considered in Chapter 5. This section is instead concerned with the provision states should make to avoid legal challenge when reforming prior rights.

It might be argued that as flowing water is *res communes*, and hence cannot be owned, then there is no property to deprive. The European Convention on Human Rights gives broad protection to 'possessions',[46] which will include the right to abstract and use water whether or not that is a right of ownership as such.[47] The state does have the right to control the use of property,[48] and the jurisprudence of the European Court of Human Rights applies principles of public interest and proportionality; and in relation to economic rights such as property, it gives states a high margin of appreciation in exercising this control.[49] Therefore a licensing regime in itself would not be a breach, but if the subsequent interference is disproportionate then compensation might be required. With adequate notice, justification and proportionality, abstraction controls should be permissible, in the public interest. In Scotland, where many abstractors took water as of right, from both surface and groundwater and including for commercial purposes, there were no challenges to the new abstraction controls.

The introduction of comprehensive abstraction licensing in England and Wales took place under the Water Resources Act 1963,[50] which granted most existing users perpetual and protected rights and did not deprive them of property. The WA2003 seeks to time-limit those perpetual rights, with notice of 12–15 years, and clear statutory grounds for restriction; the difficulties with this process will be considered below, but the Government was very mindful of the risk of challenge.[51]

In Queensland, Tan considers that the 'vested in' formula was adopted to avoid any political controversy over granting property rights in water to the Crown.[52] She suggests that there may be unresolved issues of compensation where subsequent (planned) allocations are changed or reduced, but these do not stem from the diminution in common law rights and their replacement with state control and management.[53]

In South Africa, the very particular political circumstances surrounding reform in the 1990s are unlikely to be replicated, but did provide political will. The replacement of riparian rights was not of itself an issue or an impediment to reform. Clark, using Australian examples, suggested a flexible and share-based approach would avoid challenge.[54] The South African Constitution protects property rights,[55] but also enables resource distribution on equitable grounds.[56] The slow pace of change as regards allocations may have helped to prevent challenges in the past, but these may occur in future. They are not likely to strike at the notion of state trusteeship or the fundamental power of the state to regulate, but may involve arguments about the application of the statutory criteria for compulsory relicensing, which will be analysed below.

3.3 THE AUTHORISATION OF ABSTRACTION RIGHTS

Whether an existing system gives control over water use to riparians, or to the first user of the water, or by some other means, modern water resource management requires instead a systematic allocation of water in order to make the best use of the resource. The EU WFD requires Member States to control abstractions by prior authorisation. All of the regimes studied here have introduced a comprehensive licensing system.

Two conceptual issues underpinning reform are whether a system integrates its allocation of water with other aspects of water use, and how it deals with existing rights holders. After these questions have been addressed it is possible to go on to consider the shape of the licensing regime itself, and further, whether the regime permits or encourages water trading.

The South African NWA introduces a new regime for allocation within an integrated system of water use licences.[57] These cover abstraction and also storage, impeding or diverting flow, 'stream flow reduction activities' (principally forestry, but also water-intensive non-native plants), certain specified 'controlled activities' (including irrigation with wastewater, hydro schemes and aquifer recharge),[58] discharges of waste or wastewater into a water resource, disposal of waste affecting a water resource, altering the bed, banks or characteristics of a watercourse, use of groundwater and recreational use.[59] This wide definition of water use combined with water use licences should lead to greater consistency in that the same statutory

---

[46] ECHR Art.1 Protocol 1.

[47] Thus in *Fredin v Sweden* (1991) 13 EHRR 784 the right of the property owner to exploit gravel, analogous at least with groundwater, was a property right; and in the national courts, *Catscratch Ltd No.2 v Glasgow City Licensing Board* 2002 SLT 503, so was a liquor licence.

[48] ECHR Art.1 Protocol 1 Para.2.

[49] Thus in *Jacobsson v Sweden* (1990) 12 EHRR, planning controls in general were not a deprivation; in *Fredin* revocation of a permit to exploit the gravel for environmental protection was a legitimate aim; in *Mellacher v Austria* (1990) 12 EHRR 391 the imposition of rent controls was interference, but justified in the general interest and not disproportionate.

[50] WRA 1963 c.38.   [51] DETR (1998), DETR (1999).   [52] Tan (2005).

[53] Tan (2002).

[54] Clark 'Reforming South Africa Water Legislation: Australian Examples' in FAO (1999).

[55] Constitution of South Africa s.25.

[56] Constitution of South Africa s.36.   [57] NWA ss.21–55.

[58] NWA s.37 defines controlled activities.

[59] NWA s.21 defines water use.

criteria will apply to all water activities, compared with systems where different agencies operate different control regimes, for example for abstraction, river engineering and/or pollution control.

A similar approach has been adopted in Scotland whereby all water uses, including 'activities liable to cause pollution', abstraction, impoundment and river engineering, will be controlled under the same regulations, the CAR, by the same authority, SEPA. The CAR also introduced a three-tier system. Minor and small-scale activities unlikely to cause harm to the water environment will be subject to general binding rules (GBRs). Activities that may cause cumulative damage will be subject to registration, so they can be monitored if necessary, and full water use licences will apply to activities likely to have a significant impact on the water environment, including larger abstractions. This follows the types of control mechanisms set out in the WFD;[60] a similar three-tiered approach is found in South Africa. The original proposals for the CAR envisaged that all water uses be registered, but this proved too expensive and complex, even for a small country with a sophisticated legal and administrative system. The lesson would seem to be that such comprehensiveness is unobtainable, at least in the short term.[61]

In England, the allocation of water is not currently integrated with the system of pollution control authorisations although both are regulated by the same authority, the EA. Currently abstractions and impoundments are controlled under the WRA 1991.[62] The WA2003 made certain modifications to try to address the problem of existing rights holders, but this continues to be problematic; the current Water Bill is again proposing changes to the allocation regime.

In Queensland, the QWA introduced a new system for water resource planning and allocation of water. Some aspects of rural water services are also in the QWA, but most water services law has recently been moved to separate legislation, along with control of most dams.[63] Water pollution is controlled separately under different legislation,[64] which is highly integrated across pollution control regimes; thus several different departments are involved in regulation of water.

3.3.1 Exempt and existing users and general rules

Whilst a theoretical analysis based on principles of cohesion, integrated management and comprehensive control of the resource might suggest that all water users should come within a licensing regime, this may not always be the best approach. Licensing is particular to an activity and it may be site specific or granted to a person, in which case a change of operator may need approval from the regulator. Licensing in this sense is different from an approval based for example on a code, or a general binding rule, whereby the conditions are not particular and there will not be ongoing monitoring of the specific activity. If there are many small abstractors then comprehensive licensing will be administratively expensive. Especially where these abstractions are for domestic use or subsistence agriculture, there is likely to be opposition to the rapid introduction of a scheme that licenses, and charges for, that which was previously free. At the other end of the spectrum, where private property rights are well developed, there is also likely to be resistance, and issues of compensation, when limiting or controlling these rights. There are political, social and economic advantages in taking a staged approach to reform, and dealing differently with small abstractors and larger users, whilst still accounting for cumulative impacts. Broad possibilities are to exempt some users from a licensing regime, to provide differently for existing users when controls are introduced, and to provide for some abstractions under general rules that negate the requirement to apply for an individual licence. It is commonplace to provide special protection for existing users when licences are introduced, but, as will be seen from the following analysis, it is also important to ensure that such protected rights can subsequently be reviewed and modified if necessary.

3.3.1.1   SOUTH AFRICA

In South Africa, water use is permitted by licence under the Act or authorisation under another law.[65] Use without a licence is permitted in three situations: being a continuation of an existing lawful use; use under schedule 1; or use under a general authorisation.[66] These are not site or operator specific; uses thus authorised should be registered, but do not attract recurrent fees or ongoing monitoring.

Schedule 1 uses include reasonable domestic use where the abstractor has access to the water resource; domestic use, gardening and livestock where the water resource is on the land or forms its boundary and the use is not excessive; runoff from roofs; emergency uses for human consumption and firefighting, with no restrictions as to the source; recreational use where there is access; and discharges of wastewater and runoff with the

---

[60] WFD Art.11, setting out the Programme of Measures required and making reference to prior regulation, prior authorisation and registration based on general binding rules.

[61] It has been suggested that comprehensive registration of all water users' rights will take decades, not years; Hodgson (2006) Chapter 7.

[62] WRA 1991 Part II. Part III addresses water pollution, but most discharges are now authorised under the Environmental Permitting (England and Wales) Regulations SI 2010/675; see Chapter 4.

[63] Especially, Water Supply (Safety and Reliability) (Qld) Act 2008 No.34; and see Chapter 5.

[64] Environment Protection (Qld) Act 1994 No.62, Environmental Protection (Water) Policy 2009 SL No.178; and see Chapter 4.

[65] NWA s.22.   [66] NWA s.39.

permission of the person controlling the water resource and authorised to treat the discharge.

Schedule 1 is subject to any other law controlling that use and is also subject to the Act. A CMA may, by notice and after a hearing, limit the amount of water used, where the Catchment Management Strategy identifies a need to do so,[67] which seems to strike a balance between controlling abstractions and administrative manageability.

The South African regime provides for the continuation of existing (i.e. prior) lawful uses.[68] Registration may be required by regulations,[69] enabling a fuller picture of water use to be developed. These existing uses can then be brought within the licensing regime by a voluntary or a mandatory process.

An existing user can apply for a licence, but the application may be refused on the basis of the existing entitlement.[70] However, the authorities may institute a compulsory licensing scheme, including existing users, to achieve a 'fair allocation' from a particular resource.[71] This may be done where the resource is under water stress, or to achieve equity, 'beneficial use in the public interest', efficient management, or to protect water quality.[72] The authority will then issue a proposed allocation schedule, which will reflect *inter alia* the Reserve, international obligations and existing entitlements, but also previous discrimination. After inviting and considering responses, it will issue a preliminary allocation schedule, which may be appealed, and then a final allocation. At this stage licences will replace any existing entitlement. Authorities do not need to allocate all water, but any additional water may be sold by auction or tender.

If there is a mandatory process and a licence is refused, or the amount of water granted under the licence is less than the existing use, then compensation is available if there is 'severe prejudice to the economic viability of the undertaking'. However, the calculation will disregard any reduction made for the Reserve, to rectify an over-allocation from the resource in question, or to rectify an 'unfair or disproportionate use'.[73]

The Department has developed a strategy for water reallocation for more equitable use of water, and this may involve the compulsory licensing process.[74] The intention is to redress historic disadvantage and also to promote growth and development, and progressively transfer existing uses to the licensing regime. This is inevitably contentious, as it may involve reducing existing entitlements. It is based on principles including redressing gender and racial disparities and capacity building, but has

been initiated in just three catchments. It is supported by policy and regulations providing financial assistance for 'resource poor farmers',[75] enabling grants to individual qualifying farmers or the groups within which they operate irrigation networks.

The Department is of the view that insufficient progress has been made and in the policy review document, discussed in Chapter 2, there are two proposals intended to speed up reform.[76] One is to apply the 'use it or lose it' principle to existing users, and the other is to make equity the primary factor in reallocation, instead of one factor amongst several. Soon after the publication of the policy, the Minister issued a press release to counter claims that the process was targeting farmers, but undoubtedly, as the major sectoral user, the agricultural sector will be a focus for redistribution. There was no significant challenge to the NWA when it was introduced, but cases are regularly brought to the Water Tribunal,[77] and there are likely to be individual challenges to compulsory licensing. The proposed changes should lessen the likelihood of these succeeding.

Provision is also made in South Africa for approval of some activities by a general authorisation.[78] These currently include abstractions below specified volumes; irrigation with some wastewaters, including biodegradable wastewaters from the food industry; certain discharges of wastewaters; and certain diversions and impoundments. In all cases there are restrictions both on location (by excepting certain watercourses) and on the scale of operation. The current limit for abstraction is generous, for surface water, at 150,000 m^3/annum, but there are restrictions in certain catchments and also in the coastal zone and around wetlands. There is mandatory registration for abstractions of 50 m^3/day of surface water and 10 m^3/day of groundwater. For surface water, there should be no impact on the water resource and abstraction should not be 'excessive'. These are potentially subjective qualifications, and current draft proposals will change the surface water entitlement to a maximum of 5% of the flow rate, which seems a better approach, with an annual limit of 2000 m^3 in some catchments, and reduce the limit for registration to 10 m^3/day.[79] Such general authorisations simplify the administrative regime and South Africa makes extensive use of this approach.

---

[67] NWA Sch.3.　　[68] NWA ss.32–35.

[69] NWA s.26 and Regulations Requiring that a Water Use be Registered No.1352 of 1999.

[70] NWA s.42.　　[71] NWA ss.43–48.

[72] NWA s.27. 'Efficient and beneficial use in the public interest' is also a general consideration for the issuing of licences; see further below.

[73] NWA s.22(7).　　[74] DWAF (2008).

[75] DWAF (2004a); Regulations on Financial Assistance to Resource Poor Farmers No.1036 of 2007.

[76] NWPR 2013.

[77] 'Water Tribunal Case Decisions' see http://www.dwaf.gov.za/ WaterTribunal/Cases.aspx, and further below.

[78] General Authorisations in Terms of Section 39 of the National Water Act: currently General Notice No.399 of 2004 (taking and storage); General Notice No.1199 of 2009 (diversion, impoundments, alterations to bed/ banks); General Notice No.665 of 2013 (irrigation and wastewater use); (General Authorisations). The last especially is also relevant to Chapters 4 and 5.

[79] Draft General Authorisation General Notice No.288 of 2012 (taking and storage).

3.3.1.2 ENGLAND

In England, there is a licensing regime in operation for all abstractions of water, but its effectiveness has been compromised by the establishment at the beginning of protected rights, and of licences of right for existing users.

Under the WRA 1963, licences were introduced, but with certain exceptions, including single abstractions up to 1000 gallons (approximately 4.5 m^3); abstractions used on riparian land for domestic or agricultural purposes (but not spray irrigation); and groundwater abstractions for domestic purposes. All of these abstractors were given 'protected rights';[80] authorities were under a duty not to derogate,[81] and if they did would be liable for damages for a breach of that duty.[82] More importantly, existing large users of water were granted 'licences of right',[83] which were also protected rights and were not limited by volume. Licences of right could be revoked without compensation if they had not been exercised for a period of seven years;[84] otherwise, again, compensation would be available. There was no scheme for subsequently bringing these abstractors within control. The problem with licences of right, and all rights permanently protected, is that they cannot be modified to deal with changing availability of the resource, or at least not without compensation.

These rules were then consolidated, with some minor changes under the Water Act 1989,[85] in the WRA 1991. Single abstractions of up to 5 m^3 were not controlled. Single abstractions of up to 20 m^3 needed consent but no registration or licence. Riparian occupiers could abstract up to 20 m^3/day for domestic and agricultural use, and the same volume from groundwater for domestic use, and these rights retained their protected status. The Agency could specify by notice to which parts of a landholding protected rights applied. Rules on derogation and breach of duty remained;[86] in practice, compensation is paid from the Agency's revenues from the charging scheme.

The WA2003 made some further modification. The test for licence holders is now access rather than occupation,[87] moving further away from riparianism and opening up opportunities for trading. There is a single general exemption for all abstractions of up to 20 m^3/day for any purpose;[88] that threshold can be raised or lowered, for areas or classes of water bodies, by the Secretary of State on the application of the Agency, and the Agency may be directed to apply.[89] If a user has a protected right and can no longer abstract up to the protected amount, then compensation will be available,[90] but only where the purpose of the abstraction was originally exempt and protected under the 1963 Act. Hence, the basis of the new regime is the simplest seen so far in this jurisdiction, but complicated by continuing the protection provided by earlier reforms in order to avoid extensive compensation.

The 2003 Act also seeks to better manage scarcity. There are two new types of licence, temporary and transfer licences, as well as the existing (now 'full') licences.[91] It also brings into control certain dewatering operations,[92] some water transfers,[93] and trickle irrigation.[94] In addition, the seven-year period after which a licence could be revoked on grounds of non-use is reduced to four years.[95] These so-called 'sleeper' or 'dozer' licences are problematic in England and Queensland; as we have already seen, South Africa plans a move to 'use it or lose it', whereby unused allocations will be removed. Where the resource is used occasionally but regularly, e.g., for planned crop rotation, there are fewer difficulties. However, where licences are simply held unused, then in times of shortage they will be activated, reducing availability even further. If trade is being encouraged, then these licences are likely to be sold, with the same effect, so the cancellation of such licences is an important part of reform. In England, as much as 55% of all licensed abstractions are unused.[96]

3.3.1.3 QUEENSLAND

In Queensland, since the reforms of the early nineteenth century there has been a general presumption that water uses will be licensed by the state. The QWA uses the general term 'entitlement', but there are several different types. Full water allocations are granted under an ROP within a WRP (Chapter 2), and these are severable from land and tradable.[97] Where there is no ROP, water licences will be granted;[98] these will convert to allocations when an ROP commences.[99] In addition, there are several

---

80 WRA 1963 s.26.   81 WRA 1963 s.29.   82 WRA 1963 s.50.
83 WRA 1963 ss.33–35.   84 WRA 1963 s.46
85 Water Act 1989 c.15 divested the water services providers in England and Wales (see Chapter 5) and made other substantive changes to the law on water supply, pollution control and resource management, mainly in England and Wales but also in Scotland.
86 WRA s.39.   87 WA2003 s.11.
88 WA2003 s.6, amending s.27 WRA.
89 WA2003 s.6, inserting new s.27A WRA.
90 WA2003 s.17, inserting new s.39A WRA.
91 WA2003 s.1, inserting a new s.24A WRA. Temporary licences last for 28 days or less. Transfer licences, as the name suggests, apply to the transfer of water, e.g., from one source of supply to another, or to the same source at a different point.
92 In mines and quarries, and other engineering works; WA2003 s.1, s.7. There are exceptions for emergency works.
93 WA2003 s.1, s.5. Section 5 leaves some activities exempt where carried out by, e.g., navigation and harbour boards.
94 WA2003 s.1, s.7. Controls now apply to trickle irrigation and to all other forms of irrigation not previously requiring a licence.
95 WA2003 s.17. However, these may be continued if the EA agrees; this was to address users with irregular water needs, for example agricultural irrigation of rotating crops such as potatoes.
96 DEFRA (2013a).
97 QWA Chapter 2 Part 4; see Chapter 2 on planning, and further below on trading.
98 QWA Chapter 2 Part 6.   99 QWA s.121.

different licences for the control of infrastructure.[100] As in England, certain abstractions are authorised without entitlement to owners of adjoining land (as in England, the term 'riparian' is not used in the Act). They may take water from watercourses, lakes or springs, for stock or domestic use,[101] and for camping or travelling stock.[102] Owners of land may take overland flow or subartesian water (groundwater) on that land for any purpose unless there is a restriction in a moratorium notice or a WRP, or additionally for groundwater, a regulation.[103] In addition, owners of land on which there is overland flow may take that water for stock or domestic purposes.[104] If land is subdivided it may be removed from the domestic use exemption by regulation.[105] Thus, in Queensland, the regime for exempt rights is relatively simple, applying to specified water sources and uses without complex limits by volume or historic usage, but primarily for small-scale stock and domestic use, linked to rights in land for the most part, and with powers to control the use where necessary.

3.3.1.4 SCOTLAND

In Scotland, as noted, there was no comprehensive system of abstraction controls prior to the transposition of the WFD. There is now a licensing regime, and it is also integrated to control all uses of the water environment, as in South Africa. After the CAR was first introduced in 2005, a series of orders transferred existing powers, authorising large abstractions, for public supply and hydro power.[106]

All uses at any scale will be 'authorised' (the general term used in the regulations) but not all will be licensed or registered. The intention to register every abstraction was abandoned, and now small abstractions, up to 10 m^3/day, will be controlled through GBRs.[107] The test for extending that to registration[108] or a full licence[109] will depend partly on volume and partly on the environmental consequences of the abstraction. The policy guidance indicates that abstractions of 10–50 m^3/day

will be subject to registration and over 50 m^3/day to a licence,[110] but apart from the GBR, these figures do not appear in the legislation. Where necessary on environmental grounds a higher control may be applied.[111] Thus Scotland also provides for small-scale users, within a tiered system.

3.3.2 Licensing

The essential elements of licensing schemes are a decision-maker, an application, usually in a set form and accompanied by a fee, a set of criteria by which the application will be determined, and provision for appeal. There should also be provision for time limits and/or review of the licences, public notification and representation over applications, a public register, and there may be provision for transfer.

3.3.2.1 DECISION-MAKERS

As discussed in Chapter 2, in Scotland and England the central environmental regulator controls water allocation as well as river basin planning under the WFD, whereas South Africa and Queensland both have water management and allocation under the control of a government department. In South Africa this may be devolved to the CMAs, but as seen in Chapter 2, this has not really materialised. The choice between departmental regulators and independent (or quasi-independent) agencies will also be relevant to Chapters 4 and 5.

3.3.2.2 APPLICATIONS AND PUBLICITY

Participatory structures may require the application to be notified to the public, for example by publication in the local press, or in the business paper, or both, and increasingly, electronic publication. Time periods for public responses should be specified and these must be considered rationally and reasonably, giving rise otherwise to a potential judicial or administrative review.

In England, under previous rules, all applications were advertised,[112] but the WA2003 provides instead that the Agency may advertise as prescribed, or if no regulations are prescribed, then in such a way as is calculated to bring the application to the notice of interested parties.[113] The intention is flexibility and, at least in part, to speed up licence transfers. New regulations have now been brought in to exempt certain applications from advertisement, and also to place the duty to advertise on the Agency, not the applicant.[114] Where there is already a licence, and the new application is from the same place, under the same conditions and for no greater volume, there will be no need to advertise. The Agency also has discretionary powers to exempt from

---

[100] Resource Operations Licences (ROLs) and Distribution Operations Licences (DOLs) are granted under the ROPs; QWA ss.107–108. Interim ROLs, along with interim water allocations, may be granted prior to an ROP coming into force, after which they will convert; s.167 ff. These licences will be discussed further below.

[101] QWA s.20(3). Schedule 4 provides, as does the South African legislation, that stock watering rights extend only to numbers normally pastured on a site of that size.

[102] QWA s.20(5).

[103] QWA s.20(6); specific WRPs can bring these into control in an area.

[104] QWA s.20(4). [105] QWA s.20(7).

[106] Previously, under the Water (Scotland) Act 1980 c.45 s.17 and Sch.1 and the Electricity Act 1989 c.29 s.10 and Sch.5; see Water Environment (Consequential and Savings Provisions) (Scotland) Order SSI 2006/181 and Water Environment and Water Services (Scotland) Act 2003 (Consequential Provisions and Modifications) Order SSI 2006/1054.

[107] CAR Reg.6 and Sch.3. [108] CAR Regulation 7. [109] CAR Reg.8.

[110] SEPA (2013b). [111] CAR Reg.10. [112] WRA s.37.

[113] WA2003 s.14.

[114] Water Resources (Abstraction and Impounding) Regulations SI 2006/641.

advertisement where there will be 'no appreciable effect' on the environment, or on other authorised abstractions.

In Scotland, the CAR provides for advertisement of applications where the activity has a 'significant adverse effect' on the water environment or on third parties with an interest.[115] The accompanying policy statement explains 'significant adverse effect' to mean deterioration in status of the water body, or where either conservation objectives or third party interests are compromised.[116] The extent of the obligation to advertise was contentious in the consultations over the CAR,[117] and as water use licences are the top tier of control, it might have been more appropriate if all such applications were advertised for comment. Ministers can 'call in' applications and decide them, potentially with a local inquiry, and objectors can request that this be done – but if there is no advertisement, there is no opportunity for objection in the first place.

In South Africa, the authority 'may at any stage' require the applicant to 'give suitable notice in newspapers and other media',[118] and any person who subsequently objects is entitled to notification of the decision and, upon request, written reasons.[119] The requirement to provide reasons is very important as it is a key mechanism for accountability, but the discretion here is broad.

In Queensland, where there is a finalised ROP then allocations under it will be consulted upon within the plan itself, but as seen in Chapter 2, consultation processes have been reduced in recent years. Applications for water licences (where there is no ROP) will be advertised.[120]

In all the jurisdictions therefore there is significant discretion around public consultation, which may limit the opportunity for engagement. If all applications above a certain limit, or of a certain type, were advertised, discretionary powers could apply below that limit but there would be much more clarity for stakeholders. Different arrangements for stakeholder input to water resource planning may provide another way into the process for third parties, as may rules on EIA, and separate requirements to consult other authorities (Chapter 2). Ironically, whilst the concept of IWRM makes much of participation, in every detailed control regime the trend seems to be towards increased discretion.

3.3.2.3 CRITERIA FOR DETERMINING APPLICATIONS

Each of the systems studied sets out some broad criteria in the primary legislation. In England, these are minimum flows, and not derogating from existing protected rights.[121] Where a minimum flow has not been determined, the criteria for assessing such still apply to licence applications; further, the Catchment Abstraction Management Strategies (CAMS, Chapter 2) assess flow and availability and provide information for the regulator. There are general conservation duties applying to the regulator under environmental law,[122] and to undertakers and authorities in relation to water services,[123] but no such general principles applying to allocation as such.

In Queensland, full water allocations, whether for direct abstraction or supply via a third party's infrastructure, are determined in accordance with the ROP and must be granted when the ROP comes into effect.[124] The ROP, with the WRP, should provide for ecosystems, and for existing and future water needs (Chapter 2). These requirements underpin the planning process, and allocations indirectly thereby. Resource Operations Licences (ROLs), which allow interference with flow by infrastructure, are also granted under the ROP.[125] If the related water allocation is held by another party, there must be a supply contract.[126] Where there is no ROP, and licences (or Interim ROLs) are granted, the Chief Executive must consider various factors including other users, effects on ecosystems, and effects on the resource.[127]

In South Africa the criteria are wider, partly because these are water use licences for different activities including discharges, and partly because of the political history. Authorities must consider existing uses; redressing discrimination; efficient and beneficial use; socio-economic impact of authorisation and refusal; any catchment management strategy; effects on the resource and the quality of the resource; the Reserve and international obligations.[128] The current policy review suggests that equity should be the primary factor in new authorisations, as well as for any compulsory reallocation as discussed above; and that the obligation to use the water should be strengthened.[129]

In Scotland, WEWS states certain general purposes, including preventing deterioration, promoting sustainable use, and protecting and improving the water environment.[130] The CAR then requires SEPA, before determining an application, to *inter alia* carry out a risk assessment, and assess 'what steps may be taken to ensure efficient and sustainable water use'.[131] There is a correlative general duty on all water users 'to take all reasonable steps to secure efficient and sustainable water use'.[132] SEPA must have regard to all controlled activities in the area, and may have regard to any agreements in existence[133] – for example, where various abstractors have an agreement as to timing of their abstractions.

---

[115] CAR Reg.11.   [116] Scottish Executive (2006a).
[117] Scottish Executive (2002) para.3.37.   [118] NWA s.41(3).
[119] NWA s.42.   [120] QWA s.208; so will applications for Interim ROLs.
[121] WRA ss.39–40.

[122] Environment Act 1995 s.6.
[123] WA2003 ss.81–83, and see Chapter 5.   [124] QWA s.121.
[125] QWA s.107.   [126] QWA s.121.
[127] QWA s.210, for licences; s.182, for Interim ROLs.
[128] NWA s.27.   [129] NWPR 2013.
[130] WEWS s.1, reflecting WFD Art.1.   [131] CAR Reg.15.
[132] CAR Reg.5.   [133] CAR Reg.8.

Broadly then, all these criteria are concerned with ecological impact, and the impact on current and future uses and users. Scotland has specific provision regarding efficient and sustainable use, whilst in South Africa there are several criteria including efficient and beneficial use and equity, and in Queensland the allocative plans are subject to overarching statutory principles.

3.3.2.4 CONTENT AND CONDITIONS

In South Africa, the NWA makes extensive provision on water use, beginning with an expansive definition of uses to which the Act applies.[134] There are general obligations on all water users, including a duty not to waste water.[135] The Act sets out 'general considerations' and conditions for both licences and general authorisations, and 'essential requirements' for licences. The general considerations include existing lawful uses, redress of discrimination, efficient and beneficial use in the public interest, socio-economic impact, any catchment strategy, impacts on the resource, and impacts on other users.[136] The essential requirements for licences include identifying the use, the land, the person to whom it is issued, conditions and duration.[137] The conditions may include protecting the resource, the flow and other users; management and monitoring; charges; management of wastewater and return flow; the quantity and/or percentage flow; rate of abstraction; timing and location.[138] A licensee may be required to be a member of a WUA (Chapter 2). There is special provision for stream flow reduction activities, to establish the rate of reduction and limit the detrimental effect.[139] The issue of a licence is stated not to be a guarantee of supply.[140]

In Queensland, conditions for interim allocations and licences are set out in the QWA. Interim allocations must identify the water and the abstraction location, and attach to land unless the holder has a full or interim ROL, is a public authority or a prescribed entity.[141] There may be conditions around the commencement of the abstraction; installing measuring devices; taking the water; providing and maintaining access to other entitled parties; monitoring, reporting and providing information.[142] Similar provisions are established for water licences;[143] conditions will transfer to the full allocation, modified if necessary, as authorised under the ROP.[144] The requirement to actually take the water should avoid the existence of 'sleeper' licences which hold an allocation without using it, thus depriving others who might put it to beneficial use. Interim ROLs may be mandated under a regulation,[145] or granted for future uses foreseen in the WRP or ROP.[146] Conditions for Interim ROLs

include metering and monitoring, and also prohibit any changes that would impact on other users, ecosystems, water quality or beneficial flooding.[147]

In England, the WRA required new licences to specify the quantity to be abstracted.[148] Under the WA2003, full licences must state the quantity; if for more than 12 years, they may state a minimum volume to which they can be reduced without compensation.[149] The WRA also required specification of the period(s) in which the abstraction may be made, provision for measurement, specification of works, the purpose (except for statutory undertakers) and the land to which the right attaches. The WA2003 removes the last, but requires all licences to state their purpose.[150] Under the WRA, the general effect of a licence was to create a right, and a licence also gave a defence to actions other than for negligence or breach of contract.[151] This has changed significantly; the licence will no longer be a defence to civil actions for loss or harm, and a person suffering loss or damage as a result of abstraction will be able to sue in tort for breach of statutory duty.[152] This is another measure designed to encourage responsible use by abstractors.

In Scotland, SEPA has a general power to impose conditions 'necessary or expedient for the purposes of protection of the water environment' for registered uses,[153] and a duty to do so for full licences.[154] For licences only, there must be a named person responsible for compliance. There is a further general power to impose conditions to satisfy requirements of the EU Groundwater Directive.[155] There is a formal requirement regarding surrender of a licence, to ensure that the site is left in a suitable state, which is a desirable feature. The general binding rules set out conditions for minor activities.[156] What is absent is any specification as to the types of conditions that must or may be imposed, in the general sense seen in the other jurisdictions, and indeed in the prior law in Scotland.[157] Such general conditions may be obvious, but nonetheless one would expect them to be stated.

3.3.2.5 TIME LIMITS, REVIEW AND REVOCATION

Time limits and review are fundamental. Unless states grant perpetual rights, decisions will have to be made about appropriate time periods.

In England, as most licences now are perpetual, they are not subject to periodic review, although recent reforms attempt to address some of the consequences.[158] Historic licences may be

---

[134] NWA s.21.   [135] NWA s.22.   [136] NWA s.27.   [137] NWA s.28.
[138] NWA s.29.
[139] NWA s.29; for stream flow reduction generally, see s.36. This principally refers to forestry, but the Minister can designate any use.
[140] NWA s.31.   [141] QWA s.190.   [142] QWA s.191.
[143] QWA ss. 213–214.   [144] QWA s.98.   [145] QWA s.168.
[146] QWA ss.176–178.

[147] QWA s.178; these also apply to amendments, s.182.   [148] WRA s.46.
[149] WA2003 s.19.   [150] WA2003 s.18.   [151] WRA s.43.
[152] WA2003 s.24; WRA s.48A.   [153] CAR Reg.7.   [154] CAR Reg.8.
[155] CAR Reg.9, in relation to Directive 2006/118/EC.   [156] CAR Sch.3.
[157] Control of Pollution Act 1974 c.40 s.34 gave general matters for conditions, such as place, composition, rate or temperature of discharge, and records and monitoring.
[158] WA2003 Part I, amending the WRA and discussed above.

revoked or varied on the application of the licence holder,[159] and the EA may also propose modification by notice,[160] which will be necessary if the Agency has determined that the catchment is suffering over-abstraction.[161] The EA encourages the applicant to apply voluntarily, and there is an incentive; time limits will not be added to licences modified by agreement.[162] If the licensee objects, the modification will be referred to the Secretary of State, who may hold a hearing or inquiry.[163] Whether by agreement or by notice, there may be compensation for derogation. Licences may also be revoked for non-payment of charges, or for non-use for four years.[164] If a water undertaker's licence is revoked and granted to another on grounds of efficiency, there is a new power for the agency to recover the compensation from the second licensee.[165]

Since 2012, another new power allows existing perpetual licences to be revoked or varied without compensation, to prevent 'serious damage' to waters or flora and fauna.[166] The most effective and strategic mechanism for managing over-abstraction and historic protected rights will be to use this power, in combination with the CAMS system, for identifying waters that are over-licensed, or over-abstracted. This assessment will be linked to the RBMPs and their Programmes of Measures, and in combination may also be a mechanism to assess 'serious damage'. So there are some mechanisms to address the historic issue, but the lesson for others must be to avoid granting such extensive rights in the first place.

Under the WA2003, to address these problems, new licences will be time limited. There are policy presumptions that licences will be renewed if the use is sustainable, there is a continuing requirement and there will be efficient use,[167] but Ministers repeatedly declined to put such tests on the face of the Bill, to increase flexibility for abstractors and ensure that infrastructure investment would be made.[168]

There is a policy presumption (again not stated in the Act) that new time-limited licences will be for 12 years, tying into the six-year periods under the WFD, but all licences issued in a catchment will also have a common end date, so they can all be revised at the same time under CAMS.[169] In practice therefore, licences will normally be issued firstly for between 6 and 18 years, moving to 12 years as they are reviewed together in that catchment. For new licences there will be no compensation if they are varied to protect water availability, they are only varied to a minimum amount specified in the licence, and the variation takes place after six years, with six years' notice.[170] So the new system is much better, but the historic problem is still not

resolved. The English system is also complicated by legislative style and a reluctance to state core elements of the system, even with some variability, on the face of the legislation.

At the time of writing, there is a new Water Bill for England.[171] Most of its provisions relate to water services (Chapter 5), but some parts are relevant here. In the preceding White Paper, the Government announced its intention to review the whole system of abstraction licensing.[172] In part, this is to facilitate upstream competition in water services provision, including more bulk trade.[173] The wider reform of abstractions will happen at a later stage, but the Bill includes provision to bring this within the general environmental permitting regulations.[174] The Bill removes compensation for revocation or amendment of water licences for water services providers.[175] This should make it easier to amend licences, but specifically also will ensure that, where water service providers are impacting on the quality of the water environment, the costs of this will be managed through the process for periodic review of charges and not as a general environmental cost.[176] The intention is that when the whole abstraction regime is subsequently reformed there will be no compensation for any abstractors whose rights are curtailed.[177] As this is being widely noted in Government and EA documents, there is a clear effort to ensure that effective and timely notice is being given, to forestall future challenges.

In South Africa, the relevant provisions as to duration and review are on the face of the legislation. Licences must be for a defined period not exceeding 40 years.[178] This would address the types of problems raised in England, particularly the needs of water undertakers abstracting for public supply, and for mines and quarries, both of which require long-term investment. However, there is also a mandatory review period of not more than five years, so the industry or other abstractor is not being given absolute certainty over a 40 year licence; there is a presumption, but also flexibility.

The provisions for compulsory relicensing for existing users have already been considered, but there are also general grounds for reviewing licences. When a licence is subject to its periodic review any condition other than the duration may be amended, if this is necessary to protect the resource, there is insufficient water to meet all authorised uses, or to accommodate socio-economic change and in the public interest.[179] However, such

[159] WRA s.51.   [160] WRA s.52.
[161] Under the CAMS and Restoring Sustainable Abstraction programmes; see Chapter 2, and EA (2013).
[162] EA (undated).   [163] WRA ss.53–54.   [164] WRA s.59C.
[165] WRA s.61A.   [166] WA2003 s.27.   [167] EA (2013).
[168] Hansard HC [Vol.413] Col.32.   [169] EA (2013).   [170] WRA s.61.

[171] Water Bill 2013.   [172] HM Government (2012).
[173] Water Bill 2013 especially cl.8 and cl.12; and see Chapter 5, and further below.
[174] Water Bill 2013 cl.44 and Sch.8; Environmental Permitting (England and Wales) Regulations 2010, and see Chapter 4. The Government hopes to have the new regime operating by 2022.
[175] Water Bill 2013 cl.41, amending s.61 WRA.
[176] HM Government (2013).   [177] HM Government (2012) para.2.15.
[178] NWA s.28.   [179] NWA s.49.

amendments must take place as part of a general review of all abstractors for similar uses in the same vicinity from the same resource; essentially then a general reallocation procedure, so some further protection for users against arbitrary change. Again, if there is 'severe prejudice' to the economic viability of an undertaking, then compensation may be available. There is also provision for amendment on request or by consent.[180] Suspension or revocation is available for breach or non-payment of charges. All decisions are subject to appeal to the Water Tribunal, with further appeals to the High Court on a point of law.[181] The Tribunal will also hear claims of compensation for severe prejudice. As a tribunal, its members may not be lawyers; the Act requires that they have specialist knowledge in 'law, engineering, water resource management or related fields.' It does not have jurisdiction over criminal prosecutions.

The Tribunal decisions are all made available online;[182] many of them concern procedural questions. There is some concern in the Department, both that a legal process is not the most appropriate for solving water disputes, and that, where there is an appeal, the courts may not have the necessary understanding of the water resources context to make an appropriate decision, taking into account the factors that the decision-makers would have applied.[183] The suggestion in the water policy reform paper is that it would be more appropriate to move to the general use of mediation or conciliation to resolve disputes.[184] There would still be an appeal to the courts if this route failed. The Minister currently has powers to direct that parties attempt to settle their disputes by these means; it is proposed that this would be retained and strengthened, and the Minister would also have powers to appoint an advisory panel. These types of measures seem very appropriate for water, where seeking consensus should always be the first stage, as long as they are backed by powers of compulsion.

In Queensland, most water entitlements are now provided for in a WRP; new rules allow the extension of the life of a WRP to 20 years.[185] The ROPs, which create full and tradable allocations, last for 10 years. Within a finalised ROP, amendments to conditions (in the prior licences or interim allocations) can be made if there are inconsistencies with the WRP,[186] or where provided for in the ROP itself.[187] In the Burnett Basin ROP, for example, amendments can be made to convert outstanding interim allocations to full allocations; to grant reserved allocations already in the ROP; to improve monitoring; to amend unsupplemented water allocations (i.e. those that are abstracted directly); to add or amend environmental management rules or water sharing rules.[188] These last are particularly important as there is no guarantee that the nominal volumes will actually be available.

If there is no ROP, then water will be taken by interim allocations and water licences. There is limited provision for interim allocations[189] and these will not be renewed; they are generally treated as licences. Licences apply in areas not currently subject to an ROP and will terminate in June 2111, unless an earlier date is specified in a water resource plan or ROP.[190] This was introduced in 2013, to reflect the fact that most water is now allocated under WRPs and ROPs, and the previous 10 year duration meant an unnecessary burden of renewals on licence holders.[191] Licences may be amended by the chief executive as long as the changes do not increase the volume, rate etc.; increase the area of irrigated land; increase interference with flow; change the location; or cause adverse effects on other users, ecosystems, beneficial flooding or water quality.[192] If the licensee seeks amendment it will be treated as a full application.[193]

In Scotland, there is no duration specified, even as a norm. This creates the risk that, as in England, licences will acquire characteristics of permanent property rights, the deprivation of which may then give rise to claims for compensation. In a time of global climate change and increasing pressure on resources, this seems an unnecessary regulatory risk. There is a very broad provision requiring 'periodic' review and empowering this at any time,[194] and another power to vary licences at any time, whether or not consequent to review.[195] Operators may also apply for a variation,[196] or to transfer a licence to another holder,[197] subject to SEPA's consent. There is also guidance on review.[198] SEPA reviewed all licences in order to ensure that by 2012 appropriate conditions were in place to meet the WFD objectives, by complying with the programme of measures under the RMBP. This review process also enabled identification of any environmental objectives, or measures, that require modification.[199] The RBMP process is iterative and developmental, particularly in its early stages, so flexibility is desirable;

[180] NWA s.50.   [181] NWA s.148.

[182] 'Water Tribunal Case Decisions' see http://www.dwaf.gov.za/ WaterTribunal/Cases.aspx.

[183] NWRS2 Sections 8.5.12 and 16.14.

[184] NWPR 2013 Section 3.4. These types of powers are already used under NEMA 1998 (and see Chapters 2 and 4).

[185] QWA s.52B; this will allow better use of departmental resources in the review of these Plans, especially in light of the current reforms affecting those basins within the Murray–Darling; see Chapter 2 and further below.

[186] QWA s.10.   [187] QWA s.106.

[188] DERM (2010a).   [189] QWA ss.187–199.   [190] QWA s.213.

[191] Land, Water and Other Legislation Amendment (Qld) Act 2013 No.23 s.292C.

[192] QWA s.218.   [193] QWA s.216.   [194] CAR Reg.18.

[195] CAR Regs.19–20. A variation is treated as an application as regards advertisement and determination.

[196] CAR Reg.21.   [197] CAR Reg.22.   [198] Scottish Executive (2005).

[199] As discussed in Chapter 2, the environmental objectives under the WFD, e.g., 'good' status, can be modified, or the timetable extended, for example where there is disproportionate cost; and see further Chapter 4.

however, if reviews are based around the six-year planning cycle for the RBMP then this could be stated.

SEPA has the power to suspend or revoke a licence at any time; reasons must be given, but there are no criteria stated as to the exercise of this power.[200] The policy guidance states that suspension or revocation will only be exercised 'exceptionally' where there was continuing non-compliance, with serious harm and/or widespread complaint;[201] there seems no good reason not to state these grounds in the regulations themselves.

Consistency and certainty are desirable attributes to which a modern reformed law should strive, and will help to avoid challenge. Whilst it may not be desirable to specify a single duration, in order to take account of the various needs of licence holders, where there is a norm this should be stated. It is desirable to have a longstop figure and, whether or not this is the case, licences should always be subject to periodic review. The degree of discretion given to regulators in undertaking such review may be quite wide, but the statutory criteria should be specified in law.

3.3.2.6 ENFORCEMENT

Every regime makes provision for enforcement. In South Africa, offences include using water otherwise than as permitted under the Act;[202] in Queensland, similarly, there are offences of taking water without authorisation.[203] In England, it is an offence to make abstractions without a permit;[204] in Scotland, the general offence is carrying on controlled activities without authorisation.[205] In Scotland, SEPA can issue enforcement notices for breaches of all aspects of water use.[206] In England, there is a general power of enforcement regarding abstraction and impoundment controls[207] and the regulations specify the notices available.[208] In Queensland, the Department can issue compliance notices,[209] and individuals may in some circumstances bring court proceedings themselves for enforcement orders.[210] This power is extensive; the court may award damages to the applicant or exemplary damages to the consolidated fund, and the party bringing the action need not have suffered any infringement. Similarly, in South Africa, there is a general power for responsible authorities to issue directives by notice in writing to any person in contravention of the Act.[211] In addition, individuals also have the right to raise actions or seek private prosecutions; 'any person' may seek relief in relation to breach of statutory environmental law.[212] There will be some further

general discussion of enforcement regimes in Chapter 4; the right to bring court actions in the absence of specific loss or harm is a useful tool in environmental law.

3.4 BULK WATER SUPPLIES

Bulk transfer of water may be politically and environmentally contentious. The development of national (or more likely regional) water grids may be seen to improve security of supply, but there are practical and environmental difficulties, *inter alia* the energy costs in pumping water and the undesirability of using river networks to transport water from different ecological sources. Nonetheless, there is some provision for bulk trade in each jurisdiction.

In England and especially in Scotland, there is little irrigated agriculture and much water is supplied via the water services providers (Chapter 5), which are vertically integrated, so there is relatively little provision for bulk supply. There are a number of well-established raw water transfer schemes, and over the years there have been many proposals for north–south conduits, which have usually fallen at the economic hurdle of pumping costs.[213] In England, the EA considered a regional grid in 2006 and considered it unnecessary before 2020,[214] but several years of drought have seen some renewed interest. As noted above, the intention is to bring forward upstream abstraction trading, *de facto* involving bulk transfers between water services providers.[215] Current powers to order bulk supply agreements between undertakers will be revised and strengthened;[216] and there will be an enabling power to make regulations to require undertakers to take a supply of raw water from other holders of water.[217] These will be supported by guidance and operating codes. The EA can propose a bulk supply agreement.[218]

In Scotland, where the water balance is often in surplus, there is occasional enthusiasm for exporting bulk water to our neighbours in the south. The most recent water reform here provides for an additional level of Ministerial control over bulk abstractions.[219] When first suggested by Ministers, they proposed that bulk trade with England should be encouraged; the idea had a very mixed reception, and the notes that accompanied the Bill suggested instead that the powers would be used to prevent or

---

[200] CAR Reg.25.   [201] Scottish Executive (2005).
[202] NWA s.151(1)(a).   [203] QWA s.808.   [204] WRA s.24.
[205] CAR Reg.40, Reg.5.   [206] CAR Reg.28.   [207] WRA s.216.
[208] Water Resources (Abstraction and Impounding) Regulations 2006; providing for enforcement notices, works notices and conservation notices.
[209] QWA s.780.   [210] QWA s.784.   [211] NWA s.53.
[212] NEMA 1998 gives such powers regarding breach of the principal Act; NEMA Amendment Act 2003 No.46 extends the power to breach of any

specific environmental management act, or, '*any other statutory provision concerned with the protection of the environment or the use of natural resources*' (s.6; author's italics).
[213] For a recent overview, see Pool (2013).
[214] For a discussion of these difficulties, see EA (2006a).
[215] Water Bill 2013; HM Government (2012); and see below, and Chapter 5.
[216] Water Bill s.8, inserting a new ss.40–40H into WIA.
[217] Water Bill s.12, inserting a new ss.66M–66P into WIA.
[218] WRA s.20C.   [219] Water Resources (Scotland) Act 2013 asp.5 Part 2.

control such transfers.[220] Leaving aside the economic and environmental constraints, the political difficulties of selling water outwith a country's borders are usually significant.

In South Africa and Queensland, the water services industry is vertically disaggregated, so abstraction and bulk supply are separated out from distribution to customers. In South Africa, bulk supply is provided by the Water Boards.[221] In addition, there are several bulk transfer schemes within the country and across its borders, especially the Lesotho Highlands Water Project. The Department currently operates the Water Trading Entity, and proposes that this and the Trans Caledon Tunnel Authority will be moved into a new institutional framework for National Water Resources Infrastructure.[222]

In Queensland, various water authorities may abstract and supply water, but the principal providers of bulk water are SunWater (a Government Owned Corporation, mainly supplying irrigation water) and SEQWater (the Queensland Bulk Water Authority). Much of this is relevant to Chapter 5, but the core structures and approaches will be considered here.

Queensland has seen two sets of reforms to institutional structures, including for bulk supply, in the last 10 years. As a response to a prolonged period of drought, which has driven water planning in Australia and all the states, several new bodies were established, including the Queensland Water Commission, which advised the Government on measures for scarcity and drought in South East Queensland;[223] this has now been abolished.[224] Three bulk agencies were established: LinkWater (the Queensland Bulk Water Transport Authority), WaterSecure (the Queensland Manufactured Water Authority), and SEQWater (initially, the Queensland Bulk Water Supply Authority).[225] These dealt respectively with a new water grid, recycled water and desalination, and supply of bulk water to service providers. Under a new Government, and with the drought ended, the focus is now on reducing administrative costs and the three have been amalgamated into a single body (SEQWater).[226] A regional grid was established for the heavily populated SEQ area; this will now be operated by SEQWater. Complex market rules applying between the three bulk agencies have also been abolished, and replaced by a Bulk Water Supply Code, applying between SEQ-Water and the retailers that it supplies.[227]

3.5   RAW WATER PRICING

Water pricing raises questions surrounding resource pricing and environmental externalities, also relevant to the introduction of water markets (below).[228] This section will outline the charging policies or formulas used in the jurisdictions for direct abstractors. Pricing for urban water services will be discussed in Chapter 5.

In England[229] and in Scotland,[230] the Agencies operate charging schemes that recover their administrative costs and reflect the volumes abstracted. In England, this includes an additional charge to fund compensation for removal or reduction of abstraction licences.[231] These apply to all abstractors, who will normally fund their own infrastructure. Separate arrangements are in place for the economic regulation of water services, and these fund the necessary infrastructure for public supply. In England, that regulatory system also controls access to bulk services between the water services providers.

In South Africa, the water pricing policy is currently under review.[232] It establishes charges for raw water and also for the discharge of wastewater, and includes charges for use of infrastructure managed by the Department. The revised policy is intended to be more equitable, phasing in charges for resource poor farmers whilst ensuring that commercial users pay the full cost and that those with the highest assurance of supply pay more. Prices for urban water services are managed separately. In South Africa under the general authorisations, as in Scotland for registrations, there is a single payment but no ongoing or annual fees.

In Queensland, again there are administrative fees (application and annual fees) and volumetric charges for allocations of water under the Act.[233] Where water is allocated under an ROP and subsequently traded (below) then the market will establish the price.

For bulk water, the rapidly changing policy environment in Queensland has been noted. The new supply code addresses pricing between SEQWater and the various retailers, and other users of the bulk service. The Western Corridor recycled water scheme is to be wound down, and will only be used in a future critical water shortage, and the desalination plant will be maintained in standby mode.[234] Along with the institutional reforms noted, this has enabled the current Government to offer rebates to every householder in receipt of a water bill from SEQWater.

The Queensland Government consulted in 2012 on a long-term (30 year) strategy for water, of which pricing will be one

---

220 Water Resources (Scotland) Bill SP15 Policy Memorandum.
221 DWA proposes these be reformed into Regional Water Utilities; NWRS2 Section 8.5.3; and see further Chapter 5.
222 NWRS2 Section 8.1.1. A National Water Resources Infrastructure Agency was proposed earlier, South African National Water Resources Infrastructure Agency Limited Bill 2008 B36-2008, but did not proceed.
223 Water Amendment (Qld) Act 2006 No.23.
224 SEQ Water (Restructuring) and Other Legislation Amendment (Qld) Act 2012 No.39.
225 SEQ Water (Restructuring) (Qld) Act 2007 No.58.
226 SEQ Water (Restructuring) and Other Legislation Amendment (Qld) Act 2012 No.39; and see Chapter 5.
227 Now QWA Chapter 2A Part 3; DEWS (2013).

228 See, e.g., Merret (2005).   229 EA (2012a).   230 SEPA (2013c).
231 EA (2013).   232 DWA (2013a).
233 Water Regulation 2002 SL No.74 Reg.54 and Schs.14 and 16.
234 'Bulk Water Prices' see http://www.dews.qld.gov.au/policies-initiatives/water-sector-reform/bulk-water-prices.

small part.[235] It would probably be helpful to have a period of stability, especially given the concurrent federal activities, but it is likely there will be further reform first, both on institutions and regulation.

3.6 WATER MARKETS

The focus of this chapter so far has been on the use of administrative licensing regimes to allocate water, but an alternative approach advocates the use of property rights and markets.[236] The market should redistribute water to higher value uses, for example higher value agricultural crops, or industrial use. This presumes that governments will not wish to protect certain low-value uses, such as irrigation for subsistence agriculture. It also may not provide for the environmental use of water. Water trading is still relatively unusual in global terms, and if permitted, is likely to be confined to a basin and to a sector (usually, irrigation).[237] Temporary and informal trades, especially within irrigation districts, are much more common than formal and, especially, permanent trade.[238]

Some uses of water are public goods: 'non-rivalrous', i.e. use by one person does not prevent another's use; and 'non-excludable', i.e. even those who do not pay cannot be excluded from its use. This reflects amongst economists similar ideals to the legal analysis of water as *res communes*, and not susceptible to private ownership. Public goods in water include protection of ecosystems and biodiversity, and also water quality improvements, flood management, recreational and aesthetic use. One of the difficulties then is the 'tragedy of the commons',[239] where individuals will appropriate the resource without paying for it, whenever the legal and economic system permits, and hence the resource will be both exploited and degraded. Alternatively, if resources are owned privately, then arguably the owner(s) will take responsibility for their long-term management, and hence ensure their sustainability. But private ownership does not ensure the provision of the public good aspects of water. One characteristic of public goods is that the market will not provide for them adequately, and the state will need to intervene.

A related problem is environmental externalities, particularly the costs of cleaning up pollution, and also of restoring and protecting habitats and biodiversity.[240] These are not necessarily (or even usually) reflected in the price of a resource, or in the price of a licence; they are paid for by society as a whole. In order to allocate the costs as well as the benefits of a resource to its owner, it is necessary to identify and value these environmental costs. If this could be achieved, then in theory at least it would be possible to develop perfect (or near-perfect) markets in water; external costs would be built into the price of the resource, or the price of the licence. Payments for environmental (ecosystem) services is an area of significant interest for water management (see further Chapter 4), but the valuation remains problematic.

In practice, the water resource is very rarely exposed to the full rigour of the market. Governments regulate the use and the price of water, and water is allocated within planned systems and for social reasons. Yet governments are also imperfect actors, with imperfect data, and susceptible to rent-seeking, or lobbying, or other behaviour that adversely impacts on the way they distribute resources. Some economists argue that the market would give a better indication of the true value of water to its users, and be an alternative mechanism to assess the validity of public policy choices as to the allocation of the resource.[241] This would in turn require the establishment of property rights that are sufficiently clearly defined, protected and transferable. This does not need to equate to ownership of the water resource itself. In very few countries has the resource been divested to the extent that property rights in water cannot be modified by government,[242] and even then it is still necessary for the state to regulate the market itself, providing a stable trading environment to minimise transaction costs.

Water markets can be developed within a planned system that includes an administrative allocation of water rights, as exists in all the jurisdictions studied here. The issue becomes how to balance the planning and allocation functions of the state, with the operation of markets sufficiently flexible and extensive to result in real changes in the way that the water is both valued and used.[243] There are various ways in which this can be done, and these can be seen as developmental. One possibility is to allow the transfer of water permits from one operator to another, with the approval of the regulator. This entails a generally positive and enabling approach by the state, but does not need secure and

---

[235] DEWS (2012).

[236] For a general introduction to environmental economics, discussing the underpinning concepts, see Tietenberg (2007).

[237] See, e.g., Dellapenna (2000), Dellapenna 'The Market Alternative' in Dellapenna and Gupta (2008).

[238] See, e.g., Bruns and Meinzen-Dick 'Frameworks for Water Rights: An Overview of Institutional Options' in Bruns et al. (2005); Calatrava and Garrido (2006), Grafton et al. (2010), Grafton et al. (2010a).

[239] Hardin (1968); and for refinement and critique of this concept see Ostrom et al. (1999), Ruhl et al. (2007).

[240] Tietenberg (2007) Chapter 4.

[241] See, for a pertinent discussion, Bennett 'Realising Environmental Demands in Water Markets' in Bennett (2005).

[242] California and Oregon in the USA, and Chile, though that has now been modified, were identified as having such a system; Productivity Commission (2003), Grafton et al. (2010a). The Chilean experiment did not provide adequately for environmental and social externalities, see Bauer (2004), and was subsequently modified to prevent hoarding of water rights by large companies.

[243] Freebairn 'Principles and Issues for Effective Australian Water Markets' in Bennett (2005).

severable water entitlements. This approach has been taken in England, and to a lesser extent in South Africa and Scotland. It is much more common than systems of open trading of secure and severable entitlements,[244] but the emerging markets in Australia have reached a stage where they require *inter alia* security of entitlement in order to function effectively.

3.6.1 Water trading in Scotland and South Africa: a permit-based approach

In Scotland, authorisations are transferable with the approval of SEPA; the original licensee can apply along with the potential successor, to avoid the situation where a licence is surrendered but the subsequent application refused.[245] In South Africa likewise, the NWA enables trading in specified circumstances.[246] Irrigation water may be used for other purposes by the authorised holder, or for the same purpose by another person, by the approval of the water management institution. Further, anyone with an authorisation may surrender it, or part, in order that someone else can apply for the entitlement, with the surrender conditional on the later application being successful. This seems like a useful provision addressing the principal difficulty with trade in permits, rather than secure entitlements, which is regulatory discretion. However, the Department is now proposing to remove the provisions for trading, to make more water available for equitable distribution to emerging resource-poor farmers.

3.6.2 Water trading initiatives and options in England

In England, there were some changes to improve trading options under the WA2003. This now requires access rather than occupation, but a bigger and underlying problem still leaves options at the discretion of the EA. Given the desire to reduce overall abstractions, no buyer can be certain that they will obtain the licence, but the system has been significantly streamlined and the existence of CAMS means much better data are available for prospective traders to assess whether they are likely to be successful.

As discussed above, the current round of reform proposes that in future there should be much more abstraction trading, particularly to enable upstream competition in water services. Enabling powers for some reforms are in the Water Bill, as are specific provisions for alternative suppliers of raw water to service providers and for increasing bulk supply; most recently, the Department has issued a consultation on future abstraction reforms.[247] Two possibilities are proposed. Either the current regime would be refined to be more flexible or, more radically, there would be a move to water shares rather than absolute volumes (as in Queensland and other Australian states, below). In either model, there will be a move away from any time limits on licences, but much stronger powers to vary or restrict abstractions, with a long notice period (six years is suggested). The general lessons on interference with property rights would probably suggest that these stronger powers will be less susceptible to challenge within a presumption of a time-limited licence, and the consultation responses may make this point. The consultation documents are also clearly indicating the intention not to compensate in future, well in advance, also to avoid challenge.

Most interesting in this consultation are some of the figures. There are currently some 20,000 abstraction licences and up to 55% of licensed water is not abstracted. Since 2008, the scheme for Restoring Sustainable Abstractions (Chapter 2) has affected just 121 licences so far, with 450 sites being investigated for environmental damage – but there are some 4500 exempt abstractors who should be brought into control. The intention is to focus the scheme on 305 'enhanced catchments' where the resource is under most pressure; but there will be no legislation till the next Parliament and it would be the early 2020s before the proposals are implemented, which is surely a long time in politics.

Thus the UK Government wishes to further encourage trading in future, but this will still take place in an increasingly planned system for water resource management, and where environmental impacts will be regulated and not left to the market.

3.6.3 Water trading in Australia: the policy context

In the Australian jurisdictions, water trading has been a cornerstone of federal policy since 1994, when COAG agreed a strategic framework for water reform, including trading,[248] followed by a series of initiatives tied into national competition policy.[249] A further tranche of reforms was agreed under the NWI.[250]

The NWI has overarching objectives regarding certainty (for investors and the environment) and capacity (to manage change).[251] Specific objectives include secure entitlements, transparent statutory planning, statutory provision for the environment and other public benefits, returning over-allocated or over-used systems to 'environmentally sustainable levels of extraction', removal of trade barriers, and assignation of risk regarding future availability.[252] 'Over-allocated' systems are those where the total entitlements to abstract exceed the environmentally sustainable level, whilst 'over-used' indicates that actual withdrawals are in excess of this measure.[253] 'Environmentally sustainable levels' are defined as those that would not

---

[244] Salman and Bradlow (2006) para.3.3.3.   [245] CAR Regs.25–26.
[246] NWA s.25.   [247] DEFRA (2013a).
[248] COAG (1994).   [249] COAG (1995).
[250] COAG (2004) (NWI); see also Chapters 2 and 5.   [251] NWI para.5.
[252] NWI para.23.   [253] NWI Sch.B(i).

'compromise key environmental assets, or ecosystem functions and the productive base of the resource'.[254]

As regards trading, water access entitlements must be clearly specified as a share of the 'consumptive pool' of a water resource, as determined and allocated by the relevant plan. Entitlements must be mortgageable, with public and reliable registers including any encumbrances;[255] short-term licences not amounting to secure entitlements will be discouraged. State plans must also provide for 'environmental and other public benefit outcomes', achieved by setting environmental flows, which must be as secure as entitlements, under relevant state plans. Transaction costs should be minimised[256] and barriers to trade removed.[257] There is provision for cost recovery for water services, relevant to Chapter 5.

The NWI also gives a formula for sharing risk over reductions in available water.[258] Entitlement holders will bear the risks from climate change and natural events. After 2014, under each successive water plan, where risks arise from 'new scientific knowledge', the first 3% of any reduction in the consumptive pool will be borne by the holders; between 3% and 6%, states will have a one-third share and the Commonwealth two-thirds; and above 6%, risks will be shared equally by the states and the Commonwealth. The National Water Commission (NWC) has issued a series of reports on the progress of the NWI.[259] For the Murray–Darling Basin, the NWI has been implemented *inter alia* through the Water Act 2007.

3.6.4 Water trading in the Murray–Darling: lessons for reform

The planning dimension of the reforms in the Murray–Darling Basin have been outlined in Chapter 2, along with the social conflicts that these have engendered. This is a major project; the overall budget is currently almost AUD13 billion, with almost AUD6 billion for farm support and infrastructure and more than AUD3 billion for water buybacks, over a decade. Apart from the institutional and legal dimensions, the sheer scale of trade is significant; up to 30% of basin resources will be reallocated under the current reforms.[260] Whilst it is not possible to examine the Murray–Darling in detail in this text, it will be useful to explore some of the key analyses that have been made, whilst recognising that this is an ongoing process on which much has

been and will be written, in terms of both academic comment and policy analysis.[261]

As noted in Chapter 2, the institutional and governance frameworks are complex, and have had constitutional effect. Prior to the 2007 Act, the Murray–Darling Basin had a cap on diversions, and trading rules, but these were implemented by states under identical legislation enacted in each state. The new system is driven by the Commonwealth legislation and in some ways has similarities to the implementation of Directives in the EU states. However, although the Act places obligations on states, the funding mechanism remains the most important enforcement tool. The sustainable diversion limits and purchases of water for the environment (Chapter 2) have been especially controversial, set in 2012 but coming into full effect in 2019, and the most recent Intergovernmental Agreement suggests some amendment here, to ensure there are no compulsory acquisitions of entitlements.[262] It has also been suggested that too much focus on the Basin Plan has distracted attention from other key issues, including NRM;[263] and also that the long timescale for final implementation of the Plan means that key players will no longer be there, diminishing institutional capacity.[264] There has also been some interesting analysis of the role and power of the Commonwealth Environmental Water Holder.[265] He or she will effectively control large amounts of water and, crucially, will do so long before that 2019 deadline, potentially pre-empting the states' water resource plans that should implement the Act.

A review for the Organisation for Economic Co-operation and Development (OECD) provided a list of 'lessons' for others seeking to establish water markets.[266] Suggestions included ensuring that unused allocations were revoked (otherwise they would be activated and sold); that trading is much easier to implement without the sort of priority rights seen in the USA; that externalities require separate management and incentives; that environmental water needs to be planned into the system over the medium to long term (10–15 years); that proper thought be given to the design of registers and the need to register all those with an interest in the land from which the water is being severed; and that entitlements be held individually, but that supply networks be owned by the irrigators collectively, to avoid stranded assets. Young considers that trading has been very successful, with productivity up by 13% and investors receiving 15% returns; of course, some of the productivity gains may be due to the investment in farm support, irrigation technologies etc.

---

[254] NWI Sch.B(i).   [255] NWI paras.28–33.   [256] NWI para.58.
[257] NWI para.60.
[258] NWI paras.46–51. In the most recent Intergovernmental Agreement the Commonwealth's share of risk has been slightly increased; see Intergovernmental Agreement (2013).
[259] Most recently, NWC (2011). In 2014 there will be a fuller triennial assessment.
[260] Compared to 20% in some Chilean basins, and perhaps 3% in the US states; Grafton (2010a).

[261] The NWC and the Productivity Commission have both reported extensively on activity in the Murray–Darling. See, e.g., NWC (2012), including separate chapters on activities in each state; Productivity Commission (2010).
[262] Intergovernmental Agreement (2013); this has not yet been signed by Queensland.
[263] Wallace and Ison (2011).   [264] Wallace and Ison (2011).
[265] Connell (2011).   [266] Young (2010).

(and not everyone would support a system where investors can make those returns on water). Young also suggests that the time may have come to be more specific in defining environmental water, differentiating base flow to leave in the river, and then wider uses.[267] The next section will examine how some of these lessons have been applied in Queensland.

3.6.5 Legislative provision for water trading in Queensland

The rationale for the QWA in 2000 was to achieve the policy objectives in the COAG competition policy of 1994; subsequent amendments have been made to realise the NWI and the Commonwealth Water Act 2007. The QWA sets up a framework for trading by creating secure, severable water allocations under a finalised ROP, which are then registered and are tradable.[268] Allocations may be leased, or seasonal assignments granted, but the primary purpose is to allow permanent trading.

In order to operate any trading system it is necessary to have a register of water rights. This may simply be a record of licences, as in the UK, but where secure and severable entitlements are granted it needs to be similar to a land register, with equivalent reliability and state guarantee of title, in order that buyers and lenders may satisfy themselves as to ownership and encumbrances over the right. This is an important element of a system of permanent trading.[269] Whereas all other granted water entitlements, and all draft and finalised plans produced by the public authorities, are held by the chief executive for public inspection,[270] water allocations are registered with the registrar, and the Water Allocations Register is operated as a land register.[271]

One of the issues in setting up the system in Queensland, which delayed its introduction,[272] was matching up existing securities over land to the separated water allocations. Therefore, after the draft ROP is published, interest holders may give notice to the chief executive of their intention to have that interest registered.[273] Security holders are also required to give approval to draft ROPs.

Once the ROP commences, water allocations are registered,[274] although if the allocation is under an ROL (i.e. for the infrastructure) the allocation holder must notify the registrar of the relevant supply contract; the chief executive may approve standard supply contracts.[275] The Register will contain details of the holder, a nominal volume, the location of the abstraction, the purpose, any conditions and the relevant ROP. If there is an ROL the Register will also state the ROL and the priority group. If there is no ROL, it will also state the volumetric limit, and an allocation group.[276] Holders are allocated a priority group in terms of the water sharing rules under the ROP, which will determine actual availability in times of scarcity.

To further improve the functioning of the markets, Queensland also established Distribution Operations Licences (DOLs) to distribute water through networks.[277] These are also granted under a finalised ROP, using procedures in the ROP or by application to the chief executive, and again there must be supply contracts between the users and the DOL holder. Similar provisions apply to ROLs and DOLs,[278] including conditions that they are consistent with the ROP.

Although they may be held by the same body or person, distribution networks have been separated out from the operation of the relevant dam or weir for two, related, policy reasons. The first is a general presumption under the NWI in favour of the 'unbundling' of water rights, that is, their separation into component parts to facilitate trade.[279] The second reason is to address 'stranded assets', which can arise when water is traded out of an area; previous users are no longer paying for the use of the system, and therefore the owners of the networks are left with debt but insufficient income.[280] This is also a feature of other services with similar fixed infrastructure, such as gas and railways. It is more likely to be a problem for networks than the dams themselves, so separation should make these costs more transparent. The rules in Queensland require allocation holders to pay charges to the DOL holder and to continue to pay these where distribution to those holders ceases, until the DOL holder agrees to release them from liability.[281] So again this complies with the principles in the NWI and the lessons from the OECD. Current proposals to transfer the distribution assets for irrigation channels owned by SunWater to 'local management arrangements' via irrigation boards may also be seen as unbundling, as well as divestiture to the private sector.[282]

---

[267] And see Garrick et al. (2012) for definitions of environmental water and environmental flow; and Chapter 4.

[268] QWA ss.120B–154.

[269] See, for an Australia-wide discussion holding Queensland up as a best practice model, Woolston 'Registration of Water Titles: Key Issues in Developing Systems to Underpin Water Market Development' in Bennett (2005).

[270] QWA s.1009.

[271] On principles operating under the Land Titles (Qld) Act 1994 No.11; QWA s.151.

[272] Parker (2003).   [273] QWA s.101(1)(b).   [274] QWA s.121.

[275] QWA s.122A.   [276] QWA s.127.   [277] QWA s.107A.

[278] QWA ss.108–119D.

[279] Initially, this requires the separation of water rights from land; as the market matures, it becomes desirable to separate the components of the water rights, such as rights to distribute as against rights to use. See, e.g., Campbell 'Water Trading in Australia: Some Thoughts on Future Development of Australian Water Markets' in Bennett (2005).

[280] See Roper et al. (2006).

[281] QWA s.127C. Roper et al. (2006) considered the approach adopted in Queensland to be the least economically damaging.

[282] 'Local Management Arrangements for Irrigation Channel Schemes' see http://www.lmairrigation.com.au/.

The Act also provides for dealings.[283] Changes to allocations include changes to the nominal volume, the location, the purpose, any condition, the priority group, the maximum rate, flow conditions, volumetric entitlement and the water allocation group. If there is no ROL, i.e. for direct abstractions, the change must not alter the nominal volume or increase the share of that holder within the WRP area. If there is an ROL, the change must not increase the share or the water available for supply. Allocations may be amalgamated or subdivided on approval by the chief executive. Allocations may be changed where approved under the ROP, subject to notification of any supply contract. If changes are sought that are not in the ROP, then these must be approved by the chief executive and are subject to public notification. This will be subject to compatibility with the WRP, and must be in the public interest and not significantly adversely affect other entitlement holders, ROL holders or natural ecosystems. Allocations may be forfeited on conviction of an offence under the QWA, following a show cause notice and with provision for appeal; the allocations will be sold by public auction, ballot or tender, and if there are moneys owed, the debts will be met from the proceeds. If an allocation was granted, recorded or dealt with by fraud, then the Supreme Court may make an appropriate order.

Amendments were made to the Act in 2011, to incorporate the compensation provisions and some other aspects of the Water Act 2007.[284] In the original Act, compensation was available to holders of allocations where the value of the allocation was reduced within the 10 year term of the WRP. Four of the WRP areas in Queensland are also in the Murray–Darling,[285] and for these areas only, the risk-sharing provisions from the NWI, as implemented in the Water Act 2007, are applicable in Queensland. In addition, as discussed in Chapter 2, in the Upper Condamine, a water management plan enables trading arrangements and buyback for groundwater.

Thus the trading system in Queensland is heavily plan-based. Following Commonwealth policy, trading takes place firstly in areas of the greatest water stress, that is, the same areas in which ROPs have been finalised. Trading is not an alternative to the planned management of resources and the plans will establish the protected environmental flows, and the water left for consumptive use, as well as water sharing rules and a priority order for uses and users. The current provisions of the QWA do address some of the most problematic issues, namely, the rights of security holders and the management of stranded assets.

The Australian model generally still expects the state to provide for many of the public good elements of water, and indeed to manage the structural adjustments necessary to manage the social effects of trading water, away from low-value agricultural uses which nonetheless support rural communities. Recent comparative work by Grafton across several jurisdictions sought to develop a system of benchmarking by which water markets could be compared.[286] A series of indicators, within the IWRM principles of equity, efficiency and environmental sustainability, were developed and tested in basins in jurisdictions including the western USA, the Murray–Darling, South Africa, Chile and China. Unsurprisingly, the Murray–Darling emerges as one of the most advanced and relatively successful. The report notes in conclusion that it will be preferable for there to be a close link between water planning and water trading, to ensure public good outcomes; and that where a system prioritises one of the three elements (such as social equity) the trading regime should be designed to support that, not to work against it. Although that conclusion was made in the context of South Africa, it could also apply to Australia. In a seminal article still of great relevance, Godden suggested that whilst originally the NRM model, and other aspects of Commonwealth policy, were designed to achieve (primarily environmental) objectives, the market mechanisms and trading imperatives became goals in themselves, rather than means to an end.[287] This book finds some support for that hypothesis. But it is also arguable that the Australian model is finding ways to meet the environmental objectives; what is lacking is a clear policy on addressing the social consequences.

3.7 CONCLUSIONS

The allocation of rights to abstract and use water is central to reform of water law. Where in the past water may have been allocated by customary rules, or to landowners, or to the first users, most modern water laws, including all those reviewed here, establish permitting systems. In those states that recognise private rights in water resources, pre-existing rights must be analysed and taken account of in reform proposals. These can be accommodated, either transitionally or permanently, by making appropriate provision for prior use, especially domestic and small-scale subsistence use. With a long lead time and ample warning it should not be necessary to compensate existing rights holders, especially if principles of public ownership or trusteeship, as appropriate to the legal system, are clearly established. Whilst these need not be explicit, it may be conceptually desirable and has the benefit of clarity, and of setting a marker as to

[283] QWA s.128–140.

[284] Waste Reduction and Recycling (Qld) Act 2011 No.31, Part 9, inserting new ss.986A–986J into the QWA.

[285] 'Water Resource Planning' see http://www.dnrm.qld.gov.au/water/catchments-planning/qmdb/water-resource-planning.

[286] Grafton et al. (2010a).   [287] Godden (2005).

the state's role and the primacy of the public good nature of the resource. Most important is to establish a system where water rights can be administered efficiently, including their periodic review and if necessary limitation, on clearly established statutory grounds. Because of the difficulties attendant on reforming prior rights, it is recommended that primary legislation be used for the introduction of a modern licensing regime. As noted in Chapter 2 regarding introduction of IWRM, this will also enable fuller discussion of the proposals by legislature and stakeholders and may reduce political and social difficulties with the reforms.

It is likely that a modern water law will state general principles and/or duties, for example to use water sustainably, beneficially or efficiently. It may be that broad principles are established in primary legislation setting up strategic water planning structures, which will also apply to allocation; there may be specific, narrower duties on water users. It is desirable to prioritise uses explicitly where the resource is scarce.

In terms of the structure of the reformed law, states may choose whether to control water abstractions as a separate regime, or to integrate these with other aspects of water use. The control of impoundments is usually integrated with abstraction licensing, but in Queensland, control of major dams has been moved to water supply legislation. In England, there are separate regimes, but under the same legislation and the same regulator. In South Africa and Scotland, there is an integrated licensing system for all water uses. As will also be discussed in Chapter 4, unless there are strong administrative or political reasons for taking another approach, integrated water use licences enable prioritisation of the water agenda.

Critical issues for licensing itself are the nature and extent of exempt abstractions, the length of licences and correlative powers to review or curtail, and any requirement for subsequent compensation. Clear time limits, and mechanisms for review of authorisations, will reinforce the idea that water rights operate within a planned and licensed system, under the control of the administrative authorities. In Queensland, allocation under ROPs is on a statutory basis, but planning in every jurisdiction is explicitly relevant to allocation, providing the information with which to make difficult choices and establishing structures for stakeholder engagement.

All the jurisdictions make separate provision for small abstractions and/or subsistence use, in recognition of the regulatory burden of a comprehensive regime. England uses a volumetric exemption, and Queensland recognises the rights of riparians for stock and domestic use. South Africa and Scotland both make extensive use of general rules at different scales. Whatever provision is made should be capable of integration into the full licensing regime, and be capable of variation in particular localities. No permanent rights should be granted as these will set up property rights which the state may be bound to protect, or

compensate for future interference; this will make adaptive management problematic.

For similar reasons licences should not be open-ended. Duration may be variable but should be specified, have a long-stop and periods for review. This will also prevent the acquisition of something close enough to a property right to require compensation if subsequently modified. Again, a clear expression of public ownership or trusteeship will assist. There seems no good reason not to place time periods, or the criteria for granting licences, on the face of the legislation (whether primary or secondary). If desirable and consistent with the approach of that legal system, then regulators can be given discretion in the application of the criteria, subject to appropriate appeal and review. Over-reliance on policy documents and guidance reduces certainty and clarity. In England and Queensland, the volume and complexity of the law is of concern. Users should not have to consult historic policy documents nor rely on the draftsman's explanatory notes to discover the essentials of the system.

If water rights are being reformed within the wider systems of planning and resource management discussed in Chapter 2, this will help to ensure protection of ecological quality and biodiversity, by ensuring adequate levels of environmental flows and otherwise recognising the public good elements of the water resource. If water rights are being reformed independently, in the absence of IWRM, then environmental flows must still be safeguarded to prevent ecological degradation and deterioration of water quality. In this way, allocation of water is closely linked to control of water quality and pollution, explored in the next chapter. Although IWRM is not an essential prerequisite, allocation (and reallocation) will require some of the information and data collection produced by the IWRM process.

Water trading as an allocative device is of interest in many jurisdictions, reflecting emphasis in recent years on the benefits of economic incentives as well as (but not usually instead of) regulation. It may take place with minimum regulation as a feature of a system that gives the fullest protection to perpetual property rights and the greatest emphasis on the market, or it may take place within a planned process whereby licensed rights are transferable. The former exist, but are unusual. The latter may be limited by purpose, sector or location. States may begin by facilitating transfers of licences, but if trade is to be permanent, and encouraged across sectors and at different locations, then a secure system of registering title is essential, as is provision for those with legal interests in the right, particularly security holders. Severable entitlements registered with all the formality of rights in land give greatest certainty to buyers and lenders; Queensland provides structures and models in this regard. The choice of a planned or market-based approach will be determined by a state's political environment and is beyond the scope of this book. However, the evidence suggests that it is inadvisable to create perpetual property rights in water, and a planned approach finds favour in most jurisdictions. A market approach within a planned system may achieve environmental outcomes, but it is less clear as to how to reconcile the social consequences. Whilst it is relatively easy to facilitate limited licence transfer, only in relatively sophisticated jurisdictions is there likely to be the administrative infrastructure for extensive trading at an early stage. The priority should be on establishing a planned process and a robust permitting system, and bringing users into that system.

Table 3.1 *Key findings Chapter 3: water rights and allocation*

	South Africa	Queensland	England	Scotland
Regulator	Department	Department	Agency	Agency
Vesting of resource	State; 'in trust'.	State; 'vested in'.	State/riparians (no legislative provision).	State/riparians (no legislative provision).
Integrated licences	All water uses.	No.	Abstraction and impoundment.	All water uses.
Exemptions etc.	Small users; general authorisations.	Stock and domestic.	Volume limit. (Licences of right.)	GBRs; registrations.
Tests for licences	Existing users; discrimination; efficient and beneficial use.	Accordance with ROP (ecosystems, existing and future needs).	Flow; protected rights.	Sustainable and efficient use; other users.
Duration of licences	40 years maximum; 5 year review (statutory).	10 years (statutory).	12 years presumption (policy); historic permanent licences.	Permanent licences; 6 year review (policy).
Tests for compulsory reallocation/ review and revocation	Water stress; equity; beneficial use in public interest; efficient management; water quality.	10-yearly review of ROP (property rights are then secure).	Water availability, with notice (new licences); Serious damage (protected rights).	'At any time', 'with reasons' (serious harm/ widespread complaint – policy).
Trading	Yes; permitted, limited; moving to prohibit.	Yes, extensive; severable and registered water rights; seasonal trades.	Yes, permitted; moving to extend.	Permitted, very limited.

4 Water pollution and water quality

4.1 INTRODUCTION

The policy context identified in Chapter 1 recognised the need to protect the water environment, to benefit both public and ecosystem health and maintain the resource base for sustainable development.[1] That protection can be achieved in various ways, including regulation of substances and activities that will impact on water quality.

Environmental law is well established as a branch of law, and at least some 'water lawyers' will have developed their interest from that starting point. Environmental law addresses all of the environmental media: air, land and water. It is another meta-regime in its own right, and different jurisdictions may have an integrated regime for environmental law, including water quality (which is the case in Queensland), or an integrated regime for water, covering quantity and quality (as in Scotland and South Africa).

It is arguable that, in the developed world, pollution control from point sources is a problem to which solutions are well established. Nonetheless, it remains an essential operational element of a national water law. Further, a broader water quality agenda still presents many unresolved problems, from the vexed question of diffuse pollution and consequential requirements for both behavioural change and design solutions, to the inter-relationship between water quality and quantity, to more recent debates around protecting whole ecosystems. Resource management is the agenda for the twenty-first century; pollution control is part of that. Climate change, population growth and urbanisation increase the pressures on the system in terms of quality as well as quantity. Indeed, all of these, and 'the environment' itself, can be described as 'wicked problems', deeply complex, interlinked and difficult to resolve.[2]

If a state has introduced IWRM, the frameworks examined in Chapter 2 will set the objectives to which water quality will contribute, and establish a monitoring regime that will supply important data, but there may be different institutions for pollution control. In the UK jurisdictions, there is a single environmental regulator also responsible for leading on IWRM, giving high *de facto* integration. In South Africa, the Department of Environmental Affairs (DEA) manages pollution of air and land along with biodiversity, and is in the same Ministry as DWA (Ministry of Water and Environmental Affairs). In Queensland, the Department of Environment and Heritage Protection (DEHP) manages all environmental media, separately from water planning and allocation. This chapter will set out the institutional and regulatory structures for water quality, and assess the options in the context of protection of aquatic ecosystems.

It will not look in any detail at control of other environmental media. These may also contribute to managing water quality, for example, emissions to air, causing acid rain; management of solid waste, especially leachate from landfill; and management of historic contaminated land. There is a general presumption that where there are separate control regimes, there should be formal procedures for managing the consequences of authorisations across media. The relationship with biodiversity law was noted in Chapter 2, and is especially important here; it provides reasons, and additional mechanisms, for protecting the aquatic environment. In addition, although sectoral rules are generally outwith the scope of this book, there will be some mention of the relationship between water law and the rules governing the extractive industries. Especially in Australia and South Africa, mining is a major user of water and a major polluter; the same is true for extraction of shale gas or coal seam gas.

4.2 ENVIRONMENTAL REGULATION

The regulation of the environment in the narrow sense of pollution control has traditionally utilised 'command and control', usually through licensing regimes. As in Chapter 3, 'licensing' is used here as a general term, particular to an activity, which may be site or operator specific, and will require some ongoing monitoring.

Licences make use of standards, as do codes and general rules, and these include emission limits, environmental (or ambient) quality standards (EQS), process standards and product standards. Emission limits and process standards will be used in

---

[1] See, e.g., UN (1992a) (Agenda 21) Chapter 18.
[2] The term was coined in the USA in the 1970s; for a recent review in the Australian context, see APSC (2007).

licences to achieve ambient standards.[3] As well as these, policy-makers use economic instruments, including licence fees and charges, fines, taxation, emission trading schemes and licence trading schemes; the need to account for environmental externalities, and avoid the 'tragedy of the commons', was discussed in Chapter 3. The mix between economic measures and legal tools is a complex one, requiring careful consideration in the design of the regime.[4] Where there is money available for an incentive scheme, e.g., through agricultural support, then that will be effective; but such is not always available. Another alternative would be economic measures that disincentivise, for example increasing the price of (or tax on) certain activities. Research in Europe has indicated some evidence that rises in oil prices led to increases in the cost of artificial fertilisers and caused farmers to reduce applications; and, further, that fertiliser taxes were likely to be successful and acceptable.[5]

Standards may be set uniformly, or they may be variable by the regulator, but they will be mandatory once applied. Objectives or guidelines, by comparison, are aspirational (although the terminology is not always consistent); there may be an intention that an objective will become a binding standard in a predetermined timescale. Alternatively, a guideline may be used instead of a standard, giving built-in flexibility at the cost of certainty; this is the approach in both Queensland and South Africa. The WHO suggests a step-wise approach to water quality, with a limited set of binding standards focused on the most prevalent problems in a state, that can be expanded when capacity permits; this can be supplemented by a broader set of guidelines.

These mechanisms are well suited to point sources; diffuse pollution requires a broader set of measures, although quality standards are important in identifying trends, whilst product standards, along with economic tools, can encourage behavioural change.[6] Rural land use, and sectoral rules for agriculture, are outwith the scope of the book, but given the impact on water quality, rural diffuse pollution will be addressed. The management of stormwater, which contributes to urban diffuse pollution, will be considered in Chapter 5.

4.2.1 Ecological status and an ecosystems approach

Quality standards may relate to water for particular uses, such as drinking water, but currently, the more holistic focus is on ecology, and the ability of a water body to support an adequate range of aquatic life. This development is seen in all the jurisdictions studied here, and reflects global policy agendas around taking an ecosystems approach, as developed in the CBD, and subsequently in the Millennium Ecosystem Assessment.[7] An ecosystems approach is widely described as a new paradigm,[8] and has subsequently led to discourse around ecosystem services, many of which are provided by water. Although often subject to an economic analysis, these concepts are cross-disciplinary and also useful to law and lawyers.[9] Schemes for payment for ecosystem services are an economic tool for environmental management, and these are emerging in most of the jurisdictions in relation to diffuse agricultural pollution. They are highly relevant to water safety planning and catchment protection for drinking water. Biodiversity offsetting (where, e.g., a compensatory area of habitat is required to be provided to obtain development consent, or a developer is paid to make provision for habitats or species) is a particular form of paying for ecosystem services, and is also developing. It is not possible to examine this area in any depth, and it belongs in the complex meta-regime of conservation law, but Queensland is currently introducing a Bill to rationalise several existing schemes.[10] Ruhl *et al.* find, in a US context, that payment schemes work best within regulated activities, and suggest nested planning mechanisms to develop the data necessary to include these services in decision-making. That would fit well with an IWRM approach.

An ecosystems approach is complex, and some of the data required to underpin it will be obtained through IWRM. It is a departure from traditional methods of water resource management, which concentrated on water quantity, with the principal qualitative issue being chemical quality. Environmental flows, with quantitative allocations for ecosystem protection, could be accommodated within the traditional approach; but an ecosystems approach goes much further, and is concerned not just with maintaining flow regimes but taking steps to actively improve on the much broader measure of species and habitat support. This is the water quality challenge of our day and brings together the resource assessment and planning issues, and the allocation of water to human uses and to the environment, discussed in previous chapters.

4.2.2 Enforcement and environmental justice

Whether regulation is carried out through a Government department or an independent regulator, it will be necessary to provide an appropriate set of powers (over people, premises, data, articles and substances), and a mix of enforcement tools. Sanctions may include loss of licences, financial penalties and prosecutions, and cases may be heard in ordinary or specialist courts or tribunals.

---

[3] See, for a general discussion of standards and tools intended to assist countries in developing their water quality regulations, Helmer and Hespanhol (1997).

[4] For a detailed analysis of the 'regulatory mix', which developed and advanced the arguments, see Gunningham and Grabosky (1998).

[5] Allan *et al.* (2012).   [6] Gunningham and Sinclair (2007).

[7] Convention on Biodiversity (1992); Millennium Ecosystem Assessment (2005).

[8] See, e.g., Maltby and Acreman (2011).

[9] See, for a discussion of the role of law, Ruhl *et al.* (2007).

[10] DEHP (2014); Environmental Offsets (Qld) Bill 2014.

Environmental law is complex and technical, and the consequences of poor decision-making may be borne by the most disadvantaged. There are well-known and well-established environmental policy principles, also relevant to water management, including the polluter (or user) pays; the preventive and precautionary principle; and treating problems at source.[11] These are part of the sustainable development agenda, which can be seen as the overarching policy objective, encouraging the use of economic and social tools as well as (but not instead of) regulatory instruments. Also relevant are principles of participation, including access to information and access to justice – a governance agenda. In both Australia and South Africa, such principles, including ecologically sustainable development, are prominent in the legislation. At an operational level, applications for environmental consents may be published for comment, and licences and emission data made available for inspection. In all jurisdictions, environmental assessment (Chapter 2) provides one mechanism whereby the public can obtain information and comment on environmental decision-making, but in addition specific licensing regimes should provide for information, as well as any general freedom of information rules. It is also important that environmental decisions can be challenged in a court or other appropriate forum, by affected individuals or groups; environmental NGOs can play an important role in bringing or supporting legal actions. It is not possible to examine in any detail the complex rules or case law surrounding access to justice in the case studies, but where there is a specific issue, again this will be noted.

4.3 THE EUROPEAN UNION

As noted in Chapter 2, the EU has legislated extensively on the environment and on water; early Directives included Dangerous Substances[12] and Bathing Waters.[13] In the 1980s and 1990s, there were Directives on Drinking Water Quality,[14] Urban Waste Water Treatment,[15] Nitrates[16] and Groundwater,[17] as well as Integrated Pollution Control.[18] Under the WFD, the Dangerous Substances Directive has been replaced with the Priority Substances Directive,[19] setting EU-wide EQS; the current versions of the others will remain and work with the WFD.[20] European law may use emission limits and process standards, which must be applied in Member States; for water, the EU normally sets only quality standards, and states must apply appropriate emission limits to ensure these are met downstream. The Drinking Water Quality Directive sets standards for treated drinking water, and the Urban Waste Water Treatment Directive requires the collection and treatment of sewage and biodegradable wastewater (for both, see further Chapter 5). The Bathing Waters Directive sets quality standards and management requirements for designated waters, and the Priority Substances Directive sets quality standards for specific toxic chemicals, including solvents and pesticides. Some of these are designated as hazardous, and under the WFD should be progressively phased out. The Priority Substances Directive has recently been revised, to include primarily new pesticides; in addition, certain 'emerging pollutants' (such as pharmaceuticals or personal care products) have been placed on a 'watch list' of substances that may be brought within control in future.[21] There are many thousands of chemicals in common use, including domestic use; and the interactions between them, and within the wider environment, are not fully understood.[22] In the coming decades more regulatory and policy attention will be focused on these substances, and this is also relevant to management of wastewater (Chapter 5).

As well as provisions specific to water or pollution generally, the EU is also a signatory to the Aarhus Convention,[23] and has introduced Directives on public participation in environmental matters, and on access to environmental information.[24]

It is difficult to overestimate the importance of the EU in driving forward the environmental laws, including the water laws, of its Member States. Previously, in the UK, the only statutory standard for drinking water was that it should be 'wholesome', whilst the normal provision for sewage and sludge was to dispose, untreated, at sea; now it is inconceivable that there should not be mandatory technical standards for the former, or treatment for the latter. Standards in all aspects of pollution control were set flexibly by regulators, who were often subject to regulatory capture.[25] In the UK at least, the EU is often 'blamed' by politicians for forcing environmental measures, with subsequent costs, yet at the start of the twenty-first century the measures so required are surely those that any developed government would be expected to provide.

[11] UN (1992) (Rio Declaration).   [12] Directive 1976/464/EEC.
[13] Directive 1976/160/EEC, now 2006/7/EC.
[14] Directive 1980/778/EEC, now 1998/83/EC.
[15] Directive 1991/271/EC.   [16] Directive 1991/676/EEC.
[17] Directive 1980/68/EEC, now 2006/118/EC.
[18] Directive 1996/61/EC, 2008/1/EC, now 2010/75/EU.
[19] Directive 2008/105/EC, now 2013/39/EU.   [20] WFD Art.22.

[21] Directive 2013/39/EU Arts.8b–8c.
[22] The EU is introducing the world's most complex system for managing the use of chemicals; Regulation EC/1907/2006, and see 'REACH – Registration, Evaluation, Authorisation and Restriction of Chemicals' http://ec.europa.eu/enterprise/sectors/chemicals/reach/index_en.htm.
[23] Aarhus Convention (1999).
[24] Directive 2003/4/EC; Directive 2003/35/EC.
[25] Where the regulator is insufficiently resourced and insufficiently powerful, and is unable to exert authority over the regulated. See, e.g., Hawkins (1984).

4.3.1 'Good' ecological status under the Water Framework Directive

As seen in Chapter 2, the overall objective of the WFD is 'good' ecological status. Surface water bodies are classified against a reference body as being high, good, moderate, poor and bad. The River Basin Management Plan (RBMP) sets out the Programme of Measures that will be taken during the life of that Plan to bring that water body up to good status, or alternatively to set out reasons why that standard cannot be reached. Where there are 'protected areas', higher standards may apply, and more stringent measures;[26] this will apply to areas for drinking water abstraction, and also sites protected for conservation of species and habitats.[27]

Good status requires a biological assessment of aquatic life, standards for flow and level, physico-chemical and chemical standards, and assessment of the impacts of morphological alterations to the bed and banks, such as straightening a river. In general, chemical standards have only a 'pass/fail' approach, and must be passed to achieve good status; they use a combination of emission limits and EQS. Annexes II and V are the technical annexes, and provide detail on the characterisation and classification processes, as well as a system for equilibrating the monitoring programmes carried out in different Member States.

The analysis by the Commission of the first RBMPs suggested that overall some 53% of water bodies are expected to reach good status by 2015.[28] Key issues in the first round included the adequacy of monitoring, stakeholder participation, integration of quantity and quality and climate change, better compliance with pre-existing EU environmental legislation, use of economic tools such as pricing and cost recovery, and sectoral integration, especially with agriculture.

4.3.2 Groundwater and diffuse pollution

The WFD includes groundwater in the definition of the water environment, and makes some specific requirements. States should not only 'prevent deterioration in status' of *bodies* of groundwater (author's emphasis) but also 'prevent or limit the input of pollutants' into (all) groundwater.[29] For groundwater bodies, the only classes are 'good' or 'poor', and this is determined by considering quantity (rates of abstraction against recharge), chemical quality, and effects on associated surface waters, and on dependent ecosystems such as wetlands.[30]

EU-wide quality standards apply to groundwater for nitrates and pesticides, and Member States should set 'threshold values' for substances of concern at national level.[31] Groundwater is particularly susceptible to diffuse pollution, and the WFD also requires the control of diffuse pollution as a basic measure, by prohibition or authorisation including GBRs.[32]

The Nitrates Directive requires the establishment of Nitrate Vulnerable Zones, within which there are restrictions on spreading fertiliser.[33] There are restrictions on the sale and use of pesticides,[34] and their use now requires buffer strips, but no mandatory distance is set.[35] Perhaps more importantly in the EU context, buffer strips are now required to be eligible for farm support under the Common Agricultural Policy (CAP),[36] although again distances are left to states to establish, through their implementation of the Nitrates Directive and their agricultural codes of practice. This is perhaps a missed opportunity to establish a clear minimum buffer that would be very simple to understand and apply, and could be increased if states so wished.[37] The CAP is certainly beyond the scope of this book, but where a government or other public institution provides funding for land managers (or any other sector), that is an obvious tool to direct behavioural change. As such, it is variably implemented and/or supported by other initiatives in relation to rural land use, specifically to protect the water environment. Some of these in the UK jurisdictions will be noted below. In the EU, additionally, intensive pig and poultry production beyond certain thresholds are subject to the Industrial Emissions Directive; these are treated as point sources.

4.4 ENGLAND

In England, the overarching structure and philosophy of pollution control is in the Environment Act 1995, setting out the powers and duties of the EA and its principal aim and objectives (see Chapter 2).[38] Powers of entry, etc., are found in this Act. The aim is to 'contribute towards the objective of sustainable development',[39] but the term is not defined and there is nothing comparable to the Australian statutory principle of 'ecologically sustainable development', against which more detailed provisions can be examined.

---

[26] WFD Art.4(2).

[27] Directive 1979/409/EEC, now 2009/147/EC (Birds); Directive 1992/43/EEC (Habitats).

[28] European Commission (2012b).

[29] WFD Art.4. 'Bodies of groundwater' are those capable of abstraction.

[30] WFD Annex V.

[31] Directive 2006/118/EC.   [32] WFD Art.11.

[33] Directive 1991/676/EC.   [34] Regulation EC/1107/2009.

[35] Directive 2009/128/EC Preamble, Art.11.   [36] Regulation EC/73/2009.

[37] The EU 'Water Blueprint' and associated documents identified various options for increasing cross-compliance requirements to protect water, but most of these were not adopted; European Commission (2012c).

[38] For an overview of English (and to a limited extent Scottish) environmental and water law, see Bell *et al.* (2013).

[39] Environment Act 1995 s.4.

Until recently, the principal controls on water pollution were in the WRA 1991, which established, *inter alia*, the principal offences and defences and the licensing regime for water pollution control. More recently, water discharges were included in a set of integrated Environmental Permitting Regulations (EPR);[40] it is intended that controls on abstractions will come within these regulations in the future.[41] Although local authorities have some powers under the EPR, water discharges, and discharges to groundwater, are regulated by the EA.

Recently, an Environmental Tribunal was established in England to hear appeals against some regulatory decisions.[42] Environmental appeals fall within the General Regulatory Chamber at first instance, and then a further appeal can be made to the Upper Tier Tribunal. This followed extensive lobbying by environmental and legal organisations, and academic analysis, within a wider environmental justice context.[43] However, the actual jurisdiction of the Tribunal is very narrow. In addition, the UK's record on environmental justice has some problems. As a member of the EU and a separate signatory to the Aarhus Convention, the UK jurisdictions must comply with rules on access to environmental information and environmental justice. Whilst the former are in place, implementing the relevant EU Directive,[44] the latter has been subject to much criticism. Both the Aarhus Committee and the European Courts have been concerned that judicial review is disproportionately expensive in the UK,[45] and there have been new rules to limit the costs that defendants must pay.[46]

4.4.1 Environmental Permitting Regulations

The general part of the EPR establishes the permitting system for all the relevant activities, which include waste permits and emissions from industrial activities as well as discharges to water and to groundwater. The EPR includes the regime for integrated (industrial) pollution prevention and control, and where a local authority is permitting mobile plant that discharges to water under those rules, the EA may set emission limits for such plant.[47] Emission limits are also set under the EPR for discharges to water from larger industrial processes.[48]

There is a general requirement not to 'cause or knowingly permit a water discharge activity or groundwater activity' without a permit, unless the activity is exempt.[49] The phrase to 'cause or knowingly permit' has a long history in English (and Scottish) environmental law; causation should be given its ordinary, 'common sense' meaning, and is a strict liability offence.[50] Permitting, by contrast, is a passive thing, but requires the financial and legal power to prevent the act in question.[51]

Permits are site and (usually) operator specific, and most applications are publicised for comment;[52] the consultation period is normally 30 days. If a decision is not made in the specified time (usually, four months) there is a deemed refusal. The regulator must maintain a public register of permits, with exceptions on grounds of national security and commercial confidentiality.[53] Appeals over refusals or conditions are made in the first place to the Secretary of State.[54]

Permits must be reviewed, and facilities inspected, 'periodically';[55] they may be varied on application of the operator, or by the regulator, but water discharge permits may not be varied by the regulator within four years.[56] Permits may be transferred, on a joint application by the current and prospective holders,[57] and may be revoked by the regulator.[58] The EPR also provides for standard rules, drawn up by the regulator and applying to activities of a certain type or class.[59] For water, there are standard rules for discharge of sewage below a certain volume, where there has been secondary treatment.[60]

Certain small-scale discharges are exempt. Exempt activities must be registered,[61] and then comply with the requirements of the relevant schedule. For water, the exempt activities are the cutting of riparian vegetation, and discharges from small-scale domestic sewerage (less than 5 m³/day).[62] For groundwater, there are exemptions for scientific activities (such as the input of tracers), and for discharges to soakaways from small septic tanks.[63] In all cases the activities must comply with relevant guidance.

The regulators have a series of enforcement tools, including serving enforcement notices on the operators and suspending the permit (where there is a risk of serious pollution).[64] The general part of the regulations also creates offences and penalties for their breach, and provides for remedial action by the regulators.[65] It is an offence not to have a permit when required, or to fail to

[40] Environmental Permitting (England and Wales) Regulations SI 2010/675 as amended (EPR).
[41] Water Bill 2013 cl.44.
[42] Tribunals, Courts and Enforcement Act 2007 c.15.
[43] Macrory (2002).
[44] Environmental Information Regulations SI 2004/3391; these are separate from the general regime under the Freedom of Information Act 2000 c.36.
[45] *R (Edwards & Pallikaropoulos) v Environment Agency* [2012], [2013] C-260/11; *European Commission v UK* [2013], [2014] C-530/11; Ebbesson (2013).
[46] Civil Procedure (Amendment) Rules SI 2013/262.
[47] EPR Reg.58; the IPPC system implements Directive 2010/75/EU.
[48] EPR Sch.1 para.7.

[49] EPR Reg.12.
[50] *Alphacell v Woodward* [1972] AC 824; *Empress Car Co. Ltd. v NRA* [1999] 2 AC 22.
[51] *Price v Cromack* [1975] 1 WLR 988.   [52] EPR Sch.5.
[53] EPR Regs.45–56 and Sch.24. National security is determined by the Secretary of State.
[54] EPR Reg.31.   [55] EPR Reg.34.   [56] EPR Reg.20.
[57] EPR Reg.21.   [58] EPR Regs.22–23.   [59] EPR Regs.26–30.
[60] EA (2010a).   [61] EPR Sch.2.   [62] EPR Sch.3 Part 2.
[63] EPR Sch.3 Part 3.   [64] EPR Regs.36–37.   [65] EPR Regs.38–44.

comply with a condition or a notice. There is also a series of procedural offences around making false statements or entries. There is a general defence where the act was a response to an emergency, and all reasonably practicable steps were taken and the regulators notified; and there is a separate defence for pollution from abandoned mines.[66] As well as the specified criminal penalties on conviction, the Court has the power to order remedial works. The EA issues extensive guidance on its enforcement policy and approach.[67]

In addition, there is a scheme in England for 'civil sanctions', applicable to a number of environmental offences.[68] This allows the service of various notices, and also provides for enforcement undertakings by operators, and the imposition of financial penalties by the EA. Such powers are controversial in common law jurisdictions, where there are no administrative courts and a clear separation is needed between criminal offences and civil remedies, which have a different burden of proof, so these sanctions are an alternative to prosecution. It had been expected that this scheme would be extended to the EPR, but this has not happened; the sanctions have not been widely used, and are mostly applied to waste management offences outwith the EPR. A similar scheme is being developed for Scotland, along with similar integrated permitting regulations; it is hoped it will be more effective.

Specific provision for each control regime is made in the schedules. For (surface) water,[69] a permit is required for the discharge of matter, waste or effluent into inland, coastal or relevant territorial waters (within three nautical miles of the shore); or by pipe into territorial waters beyond that point; or for the removal of vegetation, or silt or deposits from a dam or weir. In addition, a notice may be served requiring a permit for a discharge from a highway drain, or into a pond. Special provision is made for sewerage undertakers, who will not be liable for discharges of effluent in contravention of their permit, if the discharge was caused or knowingly permitted by another person; the undertaker was not bound to receive it (i.e. it was not domestic wastewater), or was bound to receive it subject to conditions that were not observed (i.e. under a trade effluent consent, Chapter 5); and the undertaker could not reasonably have been expected to prevent it. The EPR still uses the term 'controlled waters', along with definitions of inland waters, coastal waters etc., that applied prior to the WFD; under the minimalist approach taken to implementation, England has not adopted the terminology of the Directive in other aspects of the domestic law.

For groundwater, the provision is focused on meeting the requirements of EU law, so is defined accordingly.[70] Permits apply for the direct and indirect discharge of pollutants to groundwater; specified hazardous substances should not be discharged, and non-hazardous substances should be limited. There should be a site investigation for discharges to groundwater, and the regulator may serve prohibition notices, or notices requiring application for a permit. There are also special provisions here for statutory undertakers.

The EA has a groundwater protection policy, addressing point and diffuse pollution, as well as over-abstraction and other pressures.[71] The policy includes general management principles, advice on policy and legislation, and technical information, for example drilling boreholes and assessing geological suitability for various purposes.

Because the EPR includes regimes for industrial activities and waste management, and especially because of the single regulator, there is a high degree of coordination built into the institutional structure for environmental law. This should help to ensure that water is protected in other control regimes, for example leachate management when authorising landfill sites. There is less integration in terms of water licensing; abstraction and impoundments currently remain in the WRA, and are not yet in the EPR, but that will change for abstractions under proposals in the Water Bill.

4.4.2 Diffuse pollution

The Nitrates Directive is implemented by regulations, establishing Nitrate Vulnerable Zones and restricting activities in compliance with the Directive.[72] There are also separate rules on managing slurry etc., and agricultural fuel oil.[73] The Agricultural Code of Practice combines advice on meeting EU requirements for farm support and best practice advice from the Environment Agency and DEFRA,[74] and is a statutory code under the WRA,[75] which gives it evidentiary status. There is further detailed advice on cross-compliance for EU support, which requires a 2 m buffer strip for manufactured fertiliser, 10 m for organic manures, and a recommendation of 6 m as best practice.[76]

As would be expected, there have been numerous policy initiatives. The Catchment Sensitive Farming scheme has been in operation for a number of years and offers tailored support and practical solutions to 77 priority catchments via the EA and Natural England.[77] More recently, and in response to criticisms (and threatened judicial review) over implementation of the

---

[66] EPR Reg.40.   [67] EA (2011a).
[68] Environmental Civil Sanctions (England) Order SI 2010/1157.
[69] EPR Sch.22.   [70] EPR Sch.23.

[71] EA (2012b).
[72] Nitrate Pollution Prevention Regulations SI 2008/2349 as amended.
[73] Water Resources (Control of Pollution) (Silage, Slurry and Agricultural Fuel Oil) (England) Regulations SI 2010/639, 2010/1091.
[74] DEFRA/EA (2009).   [75] WRA 1991 s.73.
[76] DEFRA/RPA (2014).
[77] 'Catchment Sensitive Farming' see http://www.naturalengland.org.uk/ ourwork/farming/csf/default.aspx.

WFD,[78] a set of pilot demonstration catchments were established to manage land and water at a much smaller scale than had been done for the first RBMPs. An initial 10 catchments was broadened to 25, and now all 87 of the English WFD catchments have a partnership, though at different stages of development.[79]

Given the high proportion of water use delivered through the water utilities, as also seen in relation to water planning in Chapter 2, and given the vertically aggregated nature of the water services industry in all the UK jurisdictions, the water services providers should have a role in catchment protection. Any improvement in upstream water quality should minimise the need for treatment downstream, and several of the English companies have initiatives here, essentially making payments for ecosystem services.[80]

4.4.3 Water quality priorities: ecological status and the WFD

As the environmental paradigm becomes more holistic, with the emphasis on ecological health, implementation of the WFD is reflective of this agenda. In all the UK jurisdictions, when it comes to the technical detail of the standards and conditions that will define 'good' (and the other status classes), there has been a shared approach, with standards developed through the UK Technical Advisory Group (UKTAG) and then issued to the regulators in the form of directions. These bind the regulators, who in turn should apply them when reviewing relevant permits or licences.[81] They include, broadly, a typology of surface waters; biological standards for aquatic life; chemical and physico-chemical standards; limits for flows and levels; limits for morphological alterations; and national threshold values for groundwater.[82] It is interesting that, whereas in the early stages Scotland took a very different approach to the WFD, at the end of the process, when developing the technical parameters that actually define 'good' status, there is essentially UK-wide practice.

England is relatively densely populated, with an industrial legacy, so it is perhaps unsurprising that the results from the first RBMPs in most English RBDs were quite poor. Of the nine wholly or mainly English RBDs, only Northumbria scored more than 35% of surface water bodies at good or better status;[83] the average is around 26%, and, as noted in Chapter 2, the Government recognises that further steps will be necessary. Primary pressures in England reflect the wider EU context, being diffuse pollution, morphology and point source pollution. The Commission's analysis suggested that in England and Wales there was little evidence of the measures that would be taken in subsequent RBMPs in order to improve conditions. Despite generally good monitoring, high levels of uncertainty were used to justify postponing decisions on appropriate measures. There was also a lack of mandatory measures for agriculture (see further below), and cost recovery did not extend beyond urban water services. England will have to make significant progress in the second round.

4.5 SCOTLAND

The single environmental regulator in Scotland is SEPA, which was also set up under the Environment Act 1995 and has broad regulatory functions for almost all aspects of pollution control. SEPA traditionally had fewer strategic functions, particularly regarding water, than the EA, and reflecting this, does not have a free-standing principal aim regarding sustainable development, but only a duty to 'have regard to' whatever guidance the Ministers may issue in this respect.[84]

As already discussed, Scotland chose to implement the WFD by bringing in a whole new scheme for the management of water under WEWS and CAR, integrating all uses of water in a tiered and proportionate system involving general rules, registrations and full licences. Thus there is a high degree of integration within the water agenda. The WEWS Act and implementation of the WFD was addressed in Chapter 2, and much of the general CAR provision in Chapter 3, so the analysis here will seek to avoid repetition.

Whilst the CAR provides for transposition of the groundwater directives, which are fully integrated, and also for major abstractions for hydro schemes and domestic supply, other areas of environmental law are relevant to water, especially waste management,[85] and integrated pollution prevention and control.[86] As in England, a single regulator *de facto* provides some integration; in addition, there is specific provision in the CAR to ensure that appropriate conditions are set to protect the water environment.[87]

---

[78] WWF (2011); Angling Trust (2011).

[79] 'Catchment Based Approach' see http://www.environment-agency.gov.uk/research/planning/131506.aspx.

[80] See, e.g., the work of the West Country Rivers Trust and South West Water, 'Upstream Thinking' http://www.wrt.org.uk/; or the RSPB and United Utilities 'SCaMP' project http://www.rspb.org.uk/ourwork/projects/details/218780-scamp-sustainable-catchment-management-programme; Wessex Water, Yorkshire Water and Northumbrian Water also have major catchment programmes in operation.

[81] River Basin Districts Typology, Standards and Groundwater Threshold Values (Water Framework Directive) (England and Wales) Directions 2009.

[82] These standards etc. are currently under review for the RBMP2s; see 'UKTAG WFD' http://www.wfduk.org/.

[83] European Commission (2012d).   [84] Environment Act 1995 s.31.

[85] Environment Act 1990 Part II; Waste Management Licensing (Scotland) Regulations SSI 2011/228.

[86] Pollution Prevention and Control (Scotland) Regulations SSI 2012/360.

[87] CAR Sch.10.

A Bill currently before the Scottish Parliament will in time reform the whole system of environmental law in Scotland.[88] The intention is to bring in an integrated permitting system similar to the EPR in England, but more extensive. This will involve a tiered authorisation regime similar to that currently in the CAR; will give new enforcement powers to SEPA including civil sanctions; and will give SEPA (and many other regulators) a new (and contentious) principal aim of contributing to 'sustainable economic growth'. It is not possible to examine this Bill, or its policy context, in depth, but it moves away from a single integrated control regime for water, which this book generally supports as facilitating prioritisation of a water agenda. Alternatively, it can be seen as adopting many elements of the CAR in design of the new regime, and therefore as indicative of the success of the regulations.

In Scotland there is no environmental court or tribunal, although there are specialist prosecutors for environmental law. SEPA has powers of entry under the CAR and also under the Environment Act. As property law and private law have always been separate in Scotland, and environmental and water law are now devolved, it is unsurprising that the rules on environmental liability at common law, and judicial review, are different from England. The former is outwith the scope of this book, but the latter should be noted in terms of the environmental justice debate. Historically, the Scottish courts have been even less inclined than in England to grant standing for review, though the case law here has recently become a little more open.[89] Similarly, costs for judicial review are typically high, though as in England (in response to action by the European Commission) there has been some progress recently to limit these.[90] As in England, there are both general Freedom of Information laws and specific environmental information regulations.[91]

4.5.1 Controlled Activities Regulations

The CAR applies to the controlled activities as defined in WEWS, and also to direct or indirect discharge of certain substances to groundwater,[92] and 'any other activity which directly or indirectly has or is likely to have a significant adverse impact on the water environment'. As discussed in Chapter 3, effects will be 'significant' (only) if they affect the status of a water

body, or any conservation objectives (at least in relation to the protected areas under EU law); the bar is set quite high.

'Authorisation' is the general term for all the tiers of control. There is also the general duty regarding sustainable and efficient water use, though this seems more applicable to abstraction than to discharges;[93] that might also be true of the requirement to consider other authorised users when granting a licence.[94] The general rules are specified in the CAR; for registrations and licences, the accompanying guidance establishes the activities within each tier, but importantly, to ensure the appropriate level of control, SEPA can bring a specific activity into a higher tier.[95]

The general rules do not provide for many direct discharges, other than those via surface water drainage systems (Chapter 5). They do provide for rural diffuse pollution (below). Discharges of cooling water require registration, as do soakaways, septic tanks and other small sewerage facilities, but most discharges will require a licence.

The provisions for review, variation, etc. under the CAR have been examined in Chapter 3, along with the process for applications, and the general powers of enforcement. In an important recent case, and contrary to similar cases in England, the Scottish courts upheld the powers to impose conditions for remediation (of an abandoned opencast coal mine) on surrender of a CAR licence, even where the company causing the harm had gone into liquidation; the liquidators could not abandon the land or the obligations.[96]

In the UK jurisdictions, EIA is generally conducted within the planning framework, but insofar as it applies to some water use activities that do not require planning permission, a decision was made to build specific EIA requirements into the CAR. These will now apply to all applications where there is likely to be significant adverse impact;[97] though, as seen above, that is quite a narrow test.

There is a general offence of carrying on, or causing or permitting, controlled activities without authorisation,[98] and a series of procedural offences.[99] In comparison to England, and the prior law, there is no need to show knowledge for permitting an offence. In the Bill currently before Parliament, there will be a new hierarchy of offences, and in future the most serious offences are likely to be fault-based, whilst others will be dealt with by some form of regulatory penalty, but the final scheme for offences will come via regulations and it is not yet

---

[88] Regulatory Reform (Scotland) Bill 2013 SP Bill 26.

[89] *AXA General Insurance Limited and Others (Appellants) v The Lord Advocate and Others (Respondents)* (Scotland) [2011] UKSC 46; *Walton (Appellant) v The Scottish Ministers (Respondent)* (Scotland) [2012] UKSC 44.

[90] Act of Sederunt (Rules of the Court of Session Amendment) (Protective Expenses Orders in Environmental Appeals and Judicial Reviews) SSI 2013/81.

[91] Environmental Information (Scotland) Regulations SSI 2004/520; Freedom of Information (Scotland) Act 2002 asp.13.

[92] CAR Reg.3.

[93] CAR Reg.5.   [94] CAR Reg.9.   [95] CAR Reg.10.

[96] *The Scottish Environment Protection Agency and Others (Reclaimers)* [2013] CSIH 108.

[97] CAR Reg.11.

[98] CAR Reg.44; this would be a breach of the general requirement (to have authorisation) under Reg.4.

[99] CAR Reg.44.

clear what form it will take.[100] Defences have been reviewed recently; they include unforeseeable accidents, natural causes or *force majeure*, and a new defence for responding to civil emergencies, subject to taking all practicable steps and notifying SEPA.[101] Appeals are made to the Ministers, but in practice are heard by the Directorate for Planning and Environment Appeals.[102]

4.5.2 Diffuse pollution

In Scotland the regime is similar to England, with one novel exception. There are regulations implementing the Nitrates Directive,[103] for agricultural storage,[104] and for oil storage generally.[105] There is a similar complex structure for farm support, *inter alia* to meet EU cross-compliance rules.[106] Again there is a statutory Code of Practice for agriculture,[107] and a separate (UK-wide) code for forestry,[108] as well as a Groundwater Policy.[109] As in England, the groundwater policy applies to abstraction and both direct and indirect discharges, and it addresses links to land use planning and various special uses of land, from disposal of sheep dip to green burial sites.

Scotland has recently introduced new rules for the management of diffuse rural pollution, in the context of the WFD but going significantly beyond the requirements of the Nitrates Directive.[110] Following lengthy consultation, the Government introduced a new set of GBRs into the CAR.[111]

The new rules affect the use of fertilisers, management of livestock, land cultivation, discharge of surface water, construction of waterbound roads, pesticides and sheep dipping facilities. Many of the GBRs are concerned with setting appropriate distances from watercourses or from groundwater, including a general prohibition on cultivation within 2 m of a surface water or wetland (therefore, a mandatory buffer strip) and a duty to prevent 'significant erosion or poaching of land' by livestock within 5 m of a watercourse[112] (but not an outright prohibition on allowing cattle, which might have been clearer). There is a general obligation to cultivate land in a way that 'minimises the risk of pollution to the water environment'.[113] The most recent revisions bring new restrictions on pesticides to meet new EU requirements, and these refer to buffer zones, but with no distance prescribed and only as one means of preventing runoff.[114]

It is still relatively unusual to have legislative control of agricultural activity, as distinct from policy initiatives and financial incentives, but it is an emerging trend. The Nitrates Directive does this across the EU for vulnerable zones, and there are also developments in Queensland (below). There is a tension between environmental protection and rural land management, with a prevalent view that agricultural behaviour is difficult to regulate and should therefore be incentivised. SEPA has a Diffuse Pollution Management Advisory Group, and has been concentrating enforcement (and education) efforts in a set of Priority Catchments, where WFD requirements are likely to need significant change in land management.[115] Significant enforcement effort has included farm walks, meetings and workshops; it is too early to judge success, but there is evidence of non-compliance all over the Scottish countryside. Perhaps one or two high profile prosecutions would assist with dissemination – if the prosecutor judged these to be in the public interest.

SEPA's Priority Catchments are complemented by a new scheme operated by Scottish Water (SW, the water services provider), in catchments where improvements in land management could reduce downstream drinking water treatment costs.[116] In addition, SW has new powers to take samples etc., and enter agreements with land managers,[117] and it is likely that in the next price review period (see Chapter 5) there will be further financial provision for SW to develop these sorts of measures. This is an incentive scheme, complementing SEPA's work and providing small-scale funding for activities 'beyond compliance'. As with some of the work being done by English water companies, it can be seen as a scheme of payments for ecosystem services. Whilst financial incentives are likely to be welcomed by land managers, resources are not always available to make this the primary approach to the diffuse pollution problem. Without significant financial incentives, recent work in the USA has suggested that voluntary measures are not as effective as mandatory requirements,[118] and the regulatory approach taken in Scotland may well be part of the policy mix in many countries in the future.

[100] Regulatory Reform (Scotland) Bill 2013. [101] CAR Reg.48.

[102] CAR Regs.46–49 and Sch.9.

[103] Action Programme for Nitrate Vulnerable Zones (Scotland) Regulations SSI 2008/298.

[104] Silage, Slurry and Agricultural Fuel Oil (Scotland) Regulations SSI 2003/531.

[105] Water Environment (Oil Storage) (Scotland) Regulations SSI 2006/133.

[106] 'Farming Grants, Subsidies and Services' see http://www.scotland.gov.uk/Topics/farmingrural/Agriculture/grants.

[107] Scottish Executive (2005a). [108] Forestry Commission (2011).

[109] SEPA (2009b).

[110] Water Environment (Diffuse Pollution) (Scotland) Regulations SSI 2008/54, now incorporated into CAR Sch.3.

[111] On the consultations, the political and scientific debates and the different draft rules proposed, see Hendry and Reeves (2012).

[112] CAR Sch.3 Rule 19.

[113] CAR Sch.3 Rule 20. [114] CAR Sch.3 Rule 23.

[115] DPMAG (undated). [116] Scottish Water (2009a).

[117] Water Resources (Scotland) Act 2013 s.31.

[118] US Government Accountability Office (2013).

4.5.3 Water quality priorities: ecological status and the WFD

In Scotland, as in England, water quality priorities are driven by the WFD. Again, technical standards have been developed through UKTAG and issued by Directions,[119] and are again being updated.[120]

The country report from the Commission analysing the first RBMPs relates to the whole UK, but it does make some comment on the Scotland and Solway Tweed basins;[121] further analysis is found in the consultation on water management issues, preparing for the RBMP2s.[122] In the Scotland RBD, 67% of surface water bodies are at good or high status, and it has by far the highest percentage of high status waters in the UK. In the Solway Tweed, 47% are good or better. The Commission notes that in the Scottish RBDs the programmes of measures are established, at least as a starting point, for 2021 and 2027 as well as 2015, which gives a better idea of the Government's intentions. It is clear though that the measures relating to morphology for 2015 are the furthest behind target. As these are also the most expensive measures, and the ones (along with diffuse pollution) that require most buy-in from land managers, and as many of these have already been deferred to subsequent cycles, there is a real concern here. There will be some relaxation of morphological condition limits for the next round, following evidence that in some rivers a high degree of physical alteration has not necessarily meant a decrease in ecological abundance. However, given the very ambitious targets for 2027 (97% of water bodies to be at good or high status) it will be important that the measures under the RBMP2s are implemented fully and in good time.

4.6 AUSTRALIA

As with other chapters, in relation to water quality it is worth setting out the legal and policy context from the Commonwealth, as well as the detailed provision in Queensland. Federal Australia issues policy guidance on water quality, which should then be applied by the states; usually this is developed along with the New Zealand Government through various Ministerial initiatives. The National Water Quality Management Strategy (NWQMS) includes a suite of general and specific guidelines for managing water quality, covering *inter alia* aquatic ecosystems, primary industry, recreation and drinking water, as

well as benchmarks for diffuse pollution, sewage, effluents and recycling.[123] There is also guidance on monitoring. The principal document, with some others, was last issued in 2000 and is under review; some of the supplementary documents are more recent. The system is multi-layered, at federal, state and regional/catchment levels. Essentially these are ambient quality guidelines and, as in all the jurisdictions, are expected to work with effluent (emission) standards for point sources. The Strategy adopts a number of general environmental principles, including ecologically sustainable development, integrated catchment management, best environmental practices, and polluter/user pays.

The broad water uses covered are unsurprisingly similar to the scope of the EU directives, as are elements of the assessment and the process, but the approach taken is less prescriptive. The terminology is used throughout the Australian states; firstly, the 'environmental values', or beneficial uses, are identified for a water resource (such as drinking or recreation; aquatic ecosystems are also a 'value' in this scheme). For each value there are guidelines that should trigger an investigation or response if breached; where there are multiple values then it is suggested the most stringent should apply. Biological indicators for water, sediment and biota, as well as physical and chemical guidelines, are used. At state or catchment level, management goals should be established and, with them, water quality objectives to achieve those goals and against which progress can be measured. Objectives can be tailored locally to strike a balance between socio-economic goals and ecosystem protection; specific guidelines can also be developed more locally. The Australian system also makes use of reference waters and a classification system, and water may be one of three 'conditions': 'high conservation value', 'slightly to moderately disturbed' and 'highly disturbed'. Six types of aquatic ecosystem are identified: coastal/marine, estuarine, upland rivers, lowland rivers, wetlands, lakes and reservoirs. Generally the guidelines are said to be 'conservative', i.e. developed from high reference conditions, and it is suggested that as a management tool they will be most useful for waters of the middle class, but there are also separate indicators for the different conditions. There are similarities with the WFD, where states can apply the extensions and exemptions, on grounds of cost or socio-economic benefit, and make those trade-offs explicit. Unlike the WFD there is no presumption of a mandatory classification goal; but where there is a management goal of restoring waters, this will be reflected in tailored objectives, and may be staged.

Under the NWQMS, states were to produce water quality improvement plans, and this was done in Queensland through the SEQ Healthy Waters partnership; Moreton Bay (which, like

---

[119] Scotland River Basin District (Surface Water Typology, Environmental Standards, Condition Limits and Groundwater Threshold Values) Directions 2009. Solway Tweed River Basin District (Surface Water Typology, Environmental Standards, Condition Limits and Groundwater Threshold Values) Directions 2009.

[120] Scottish Government (2013). [121] European Commission (2012d).

[122] SEPA (2013a).

[123] ANZECC/ARMCANZ (2000).

the adjacent Great Barrier Reef, is a Marine Park) was designated a Water Quality Hotspot. In Queensland, this work is being taken forward *inter alia* by Healthy Water Management Plans (below).

In addition to the extensive policy context, federal Australia also legislates for environmental protection and biodiversity. The Commonwealth Water Act 2007 and ecological improvements in the Murray–Darling are one element of this; the Basin Plan sets targets.[124] Also relevant is the Environmental Protection and Biodiversity Conservation (EPBC) Act, implementing the Biodiversity Convention and providing for environmental assessment at federal level, or under bilateral agreements with states. Water resources have recently been added to the list of matters of 'national environmental significance' under that Act.[125] Specifically, coal seam gas extraction and large coal mining activities with a significant impact on water resources have been added to the developments that require approval, and possibly environmental assessment, at federal level. However, there is a commercial, and political, tension with environmental goals. At the time of writing, the federal Government has consented to dredging at Abbot Point, in Queensland, to extend a port used for the burgeoning coal industry; and the Great Barrier Reef Marine Park Authority has given permission for depositing the soil on the reef.[126]

Whilst all the jurisdictions have numerous functioning environmental NGOs, until recently Australia had publicly funded Environmental Defenders' Offices. It is disappointing that federal funding for these has recently been withdrawn.[127]

4.7 QUEENSLAND

In Queensland, water quality and pollution control are primarily managed through the DEHP, and there is an integrated system of environmental law, but with media-specific 'policies' (which are regulatory), including for water. This is generally separate from the QWA and the structure for water abstractions, although some elements of the QWA are relevant to management of water quality.[128] Queensland (as with other Australian states) has a

zoned system for land use planning under the Sustainable Planning Act; some environmental approvals may be integrated within development consents.[129] For environmental consents, and some approvals under the QWA, one department may be the assessment manager, and another a 'referral' or 'concurrence' agency for that decision. Although this system of referrals is complex, it is explicit and therefore transparent. The Queensland Government, like the UK Government, is currently keen to pursue a deregulatory agenda and has therefore reduced the need for full licences for smaller operators in its most recent reforms, and is making more use of standard rules.[130] Queensland also has its own Water Quality Guidelines, which are more localised than the Australian guidelines.[131] They work with the Water Policy (below); where the state has not identified a parameter, the Commonwealth guidelines will apply.

4.7.1 Environmental Protection Act

The Environmental Protection Act (EPA)[132] was passed in 1994, but has been much amended; its overall objectives are framed in terms of ecologically sustainable development. The accompanying Regulation[133] and Water Policy[134] are more recent, but again are regularly updated and amended. The Act establishes certain levels of environmental harm, and requires 'environmentally relevant activities' (ERAs) to be approved.[135] It uses the general term 'authority' to cover all levels of consent, whether subject to standard rules, or site specific.[136] If the activity involves a change of use, an application for development consent must be made, and this will also be taken as an application for an environmental authority.[137] As part of the deregulatory agenda, new provisions in the Act enable the use of standard conditions and codes of practice, and allow operators to be registered with the Department, to streamline their applications.[138]

The Act provides for Environmental Impact Statements and public consultation on these.[139] As would be expected, the licensing regime provides for variation, transfer, revocation and surrender of approvals, and the last may require remediation etc. of the site.[140] Applicants, operators and recipients of notices may seek first an internal review, and then appeal to Court – usually, the

---

124 Commonwealth of Australia Water Act 2007 Basin Plan Sch.11, and see Chapter 2.
125 EPBC Act 1999 as amended, ss.24D–E. Activities affecting Ramsar wetlands, and the Great Barrier Reef, were already covered by these provisions.
126 'Abbot Point Capital Dredging Project' see generally http://www.gbrmpa.gov.au/about-us/consultation/current-proposals/abbot-point-capital-dredging-project.
127 'Australian Network of Environmental Defenders' Offices' see http://www.edo.org.au/. The Queensland EDO successfully sought review of a decision to approve a dam under the federal EPBC process without considering the impacts on the reef of diffuse agricultural pollution (below); *Minister for the Environment and Heritage v Queensland Conservation Council Inc.* [2004] FCAFC (the *Nathan Dam* case).
128 Riverine protection, QWA ss.266–273; groundwater management, Chapter 3-3A (below).

129 Sustainable Planning (Qld) Act 2009 No.36; Chapter 6 provides for the Integrated Development Assessment System.
130 DEHP (undated). 131 DEHP (2013).
132 Environmental Protection Act (Qld) 1994 Act No.62 (EPA).
133 Environmental Protection Regulation 2008 SL No.370 (EP Regulation).
134 Environmental Protection (Water) Policy 2009 SL No.178 (Water Policy).
135 EPA ss.18–19. 136 EPA ss.121–124.
137 EPA s.115. In that case the Integrated Development Assessment System under the SPA will apply; SPA Chapter 6.
138 EPA Chapter 5A, substituted by the Environmental Protection (Greentape Reduction) and Other Legislation Amendment (Qld) Act 2012 No.16.
139 EPA Chapter 3. 140 EPA Chapter 5.

Planning and Environment Court, but the Land Court deals with some aspects of mining and resource use. In terms of enforcement, regulators may serve environmental protection orders, direction notices, cost recovery and clean-up notices.[141] There is a general duty not to carry out activities likely to cause environmental harm without taking all 'reasonable and practicable' steps to mitigate the risk.[142] It is an offence to carry out an ERA without authorisation or in breach of conditions, and there are general offences of causing serious or material environmental harm, or environmental nuisance,[143] and particular offences relating to water.[144] Offences may lead to criminal prosecution and the Court may order remediation and the payment of costs.[145]

The Act makes provision for some agricultural ERAs, applicable to rural diffuse pollution (below);[146] and also special protection for activities (such as sewage treatment) in areas covered by the Wild Rivers Act.[147] This protects pristine or near-pristine catchments, which are not protected sites such as national parks, but where the usual trade-offs between development and ecology should be inclined towards the latter.

4.7.2   The Environmental Regulation

As would be expected, the Regulation provides additional detail on decision-making processes, *inter alia* on EIAs, and on the various ERAs. It specifies which ERAs are concurrence activities also requiring development consent,[148] and the relationship with local government. ERAs with particular relevance to water include water and sewage treatment, aquaculture, intensive animal husbandry and some food processing. There are additional controls for discharges to wetlands or groundwater,[149] and a list of proscribed contaminants for water.[150] The Regulation provides for fees, and certain public registers etc.[151] There is a general presumption that all environmental authorisations should protect the environmental values of the different media;[152] the environmental values for water are established through the Water Policy.

4.7.3   The Water Policy

The Policy is established by the Minister under the Act.[153] It establishes environmental values, and may establish objectives, indicators, programmes and standards. Despite the name, the

Policy has the status of a statutory instrument. It provides for environmental values, management goals, water quality guidelines and objectives, and a framework for decision-making and reporting. There is a general statement as to the various values that are protected,[154] and here Queensland refines the Commonwealth approach, using a four-fold classification; waters of high ecological value and highly disturbed waters are the same, but there is a subdivision of the middle class, 'slightly or moderately disturbed'. 'Slightly disturbed' waters have unmodified biology, but slightly modified chemical or physical characteristics. For 'moderately disturbed' waters, the biological integrity is measurably adversely affected. This precedence of the biological assessment also recalls the WFD. In addition, the following uses are specified as values, following the NWQMS: aquaculture, agriculture, recreation, industry and cultural and spiritual use.

The Policy sets out a process for the Department to establish (measurable) indicators and guidelines, and then management goals and water quality objectives, for specified waters. There should be public consultation, and there is a general provision that where an objective is lower than the guidelines, this must be to prevent unacceptable social and economic impacts, and there must still be an improvement in quality.[155] This would be a useful approach for any country using guidelines rather than standards. The Healthy Waters Management Plans are also provided for, and should include any values, objectives and guidelines as well as any environmental flow objectives or ecological outcomes in a relevant Water Resources Plan; this is a reminder of the divide in Queensland between planning for quantity and for quality, which adds to the complexity of the system. In addition, the non-statutory bodies outlined in Chapter 2 are active in water quality monitoring, especially the Healthy Waterways Partnership.

For many waters, these general provisions are overtaken by specific values and objectives established in detailed assessments of particular river systems. The rivers for which this has been done are listed in the Policy and then a separate report is available for each.[156] At this level, the policy suite is drilling down to the level of detail found in the programmes of measures under the EU RBMPs.

4.7.4   Diffuse pollution

In Queensland and Australia, as in other jurisdictions, diffuse pollution from agriculture is a potential threat to water quality. The system discussed above at both Commonwealth and state levels provides water quality objectives for agricultural pollutants, but for the most part agricultural activities are not ERAs

---

[141] EPA Chapter 7 Parts 5, 5A–C; powers of entry, etc. are in Part 9.
[142] EPA s.319.   [143] EPA Chapter 8.   [144] EPA Chapter 8 Part 3C.
[145] EPA Chapter 10.   [146] EPA Chapter 4A.
[147] EPA s.174; Wild Rivers (Qld) Act 2005 No.42.
[148] EP Regulation Sch.2.
[149] EP Regulation Chapter 4 Part 3; and specific environmental values for wetlands, Reg.81A.
[150] EP Regulation Reg.77 and Sch.9.
[151] EP Regulation Chapter 7 Part 4. Queensland also has a general Freedom of Information Law; Right to Information (Qld) Act 2009 No.13.
[152] EP Regulation Reg.51 and Sch.5.   [153] EPA s.26ff.

[154] Water Policy s.6.   [155] Water Policy s.11.
[156] Water Policy Sch.1; and see generally 'Water Publications' http://www. ehp.qld.gov.au/water/monitoring/publications.html.

and therefore there are few enforcement mechanisms as such. Keeping intensive livestock (and aquaculture) above certain thresholds is an ERA and, as in the EU and UK, is a point source.[157] Cottingham *et al.* analysed the legislative, institutional and policy environment for all sources of diffuse pollution in Queensland in 2010 and indicated a complex set of relationships, given the size of the jurisdiction and even allowing for the federal structure.[158]

Agricultural runoff has impacts on marine and coastal, as well as fresh, waters. In recognition of this, and under an agreement with the Commonwealth, Queensland introduced some mandatory rules on diffuse pollution, specifically to protect the Great Barrier Reef.[159] These create ERAs for cattle and sugarcane above certain thresholds, and require relevant farmers to produce Environmental Risk Management Plans. The management of the Reef is a matter of local, regional and global interest, with multiple overlapping conservation designations and institutions. Currently, the Queensland and Australian Governments are carrying out a strategic assessment of its management, under the Commonwealth EPBC Act,[160] but as noted, pressure for development consents remains high in Queensland. Waska and Gardner have analysed the new ERAs, which are limited to three priority catchments and to a very narrow set of activities, and suggested that a more extensive approach, applying to more activities and at smaller scales, might be more effective.[161] They are also concerned that new regulations may alienate those farmers who are already most proactive in managing their activities, which is a general issue. They suggest that a flexible approach to the structure and content of Environmental Risk Management Plans is helpful here; certainly Queensland is more flexible in this regard than the farm management plans produced in the UK to meet EU requirements. Previous general provisions for Land and Water Management Plans by farmers in the QWA have been repealed, and though one of the reasons for introducing the Wild Rivers Act was the absence of any mandatory codes on farming practice, none have so far been produced under that legislation either. Given the deregulatory focus of the current Queensland Government, that seems unlikely at present.

Aside from the Reef, management of Moreton Bay (also a Marine Park) also raises concerns about the impact of land-based pollution (rural and urban) in coastal waters. Given the very high density of population in SEQ compared to other parts of the state, the SEQ regional arrangements are especially important, and impact on the Reef and the Bay as well as inland ecology. The ecological monitoring by the Healthy Waters Partnership monitors the effect of diffuse as well as point source pollution; Waska and Gardner suggest that the programme, with its system of scorecards, may be having a negative effect on the local governments which have responsibilities, but few powers, and support and fund the partnership.

As would be expected, there is extensive best practice guidance from the Department of Agriculture, Fisheries and Forestry, especially in relation to salinity and soil management.[162] In relation to water quality, the focus is on the Reef catchments, indicating the power of a legislative, as distinct from policy, framework. Mandatory requirements at federal level led to legislation at state level, which in turn motivates regulators and decision-makers and guides the allocation of resources.[163] This in itself is a strong argument in favour of mandatory regimes. In both the UK and Queensland, environmental improvements tend to be driven by a higher power.

4.7.5 Groundwater

In Queensland there is no single overarching policy for groundwater, but there are several ways in which it can be protected, including controlling allocation, which will in turn assist in protecting related surface waters and groundwater-dependent terrestrial ecosystems. As noted in Chapter 3, under the QWA landowners may take subartesian water unless a restriction is in place, which may be in a WRP, a moratorium notice, or in the Water Regulation. WRPs can establish 'groundwater management areas' and, within them, require water licences and development permits. The Water Regulation also provides for groundwater management areas, for seasonal allocations, water sharing rules, water harvesting and metered entitlement;[164] and for stock and domestic rights in 'subartesian areas'.[165] The general power to issue moratoriums is exercised by the Minister under the QWA, to protect ecosystems or other users, and a moratorium is effective even if it conflicts with a WRP or ROP.[166] There is a specific licensing scheme for drilling boreholes.[167]

Although sectoral rules on mining and other extractive industries cannot be considered in any detail, they are major contributors to Queensland's economy and major users of water, especially underground water. They have the potential to cause significant pollution, both in their direct operation and related activities such as transport; the recent Federal decision to allow deposit of dredged materials on the Great Barrier Reef has been mentioned.

---

[157] EP Regulation Sch.2 Part 1.   [158] Cottingham *et al.* (2010).

[159] EPA Part 4A.

[160] 'Great Barrier Reef Strategic Assessment' see generally http://www.dsdip.qld.gov.au/gbr-strategic-assessment/.

[161] Waska and Gardner (2012).

[162] 'Queensland Government Sustainable Agriculture' see generally http://www.daff.qld.gov.au/environment/sustainable-agriculture.

[163] 'Queensland Government Water Quality and Water Use' see generally http://www.daff.qld.gov.au/environment/sustainable-agriculture/water-quality-and-water-use.

[164] Water Regulation Schs.4, 10, 14, 15A.   [165] Water Regulation Sch.11.

[166] QWA s.26.   [167] QWA Chapter 2 Part 10.

There is currently a high profile global debate over hydraulic fracturing ('fracking') as a method of extracting shale gas and coal seam gas; the latter is extensively used in Queensland both for domestic gas and for export as liquid natural gas. The last Government introduced a series of reforms for holders of petroleum licences, specifically to protect groundwater.[168] There is a requirement to conduct an underground water impact assessment and submit a report, including monitoring of quantity and quality of the relevant aquifer and the impact on any associated springs and their ecosystems. Before the tenure ends a final report must be submitted, and conditions can be attached that extend beyond the life of the tenure. Where several petroleum tenure holders have rights affecting the same aquifer, a Cumulative Management Area can be established. The current Government created an Office for Groundwater Impact Assessment, which manages these processes and provides reports on Cumulative Management Areas, following abolition of the Queensland Water Commission.[169]

Queensland is not alone in making provision for fracking; South Africa promulgated draft regulations,[170] whilst the UK Government has expressed its encouragement. However, Queensland's legal framework is the most advanced of the jurisdictions studied here. Assessment and reporting will be one element of the management framework for this process, but is unlikely to allay concerns or prevent protests. The longer-term impacts on water supply remain the most pressing concern.[171]

4.8 SOUTH AFRICA

In South Africa, environmental law is the responsibility of the DEA, although other departments, and levels of government, also have roles and functions.[172] The South African Constitution provides a right, albeit qualified, to a healthy environment and environmental protection, which sets the context for sectoral legislation.[173] The principal Act is NEMA 1998, and there is series of specific Acts, dealing for example with protection of biodiversity, or waste management. NEMA provides for cooperative governance and sets out core environmental management principles, including sustainable development, polluter pays, and minimising waste at source. There are also principles around environmental justice and participation, and a statement that the environment is held in public trust.[174] Subsequent amendments provided *inter alia* for environmental impact

assessments[175] and also for enforcement powers and an inspectorate.[176] The DWA can appoint inspectors[177] and specifically is a concurrence authority for waste permits and can exercise compliance powers for waste.[178] This is especially important for the management of leachate from landfill, which has the potential to contaminate groundwater. In the past DWAF managed landfills, but this seems a more sensible approach. The DWA is also to be consulted over steps taken to remediate contaminated land.[179]

NEMA makes some wide-ranging provision affecting all environmental media. There is a general duty, widely expressed, requiring 'every person' who 'causes, has caused or may cause significant pollution or degradation of the environment' to take 'reasonable measures' to prevent such pollution or degradation, or where such is authorised, to minimise its effects.[180] The environment is defined to include 'land, water and atmosphere', potentially overlapping with general provisions in the NWA.[181] The NWA itself allows for water use licence requirements to be met by other authorisations,[182] and for authorities to liaise and combine 'their respective licensing requirements into a single licence'.[183]

NEMA provides generally for high level horizontal coordination between departments, as well as vertical cooperation through different tiers of government. Relevant departments should produce environmental management plans and/or environmental implementation plans;[184] both apply to DWA. There are mechanisms for authorities to seek conciliation as well as arbitration, to resolve disputes, and any person may request a facilitator to be appointed; courts may order conciliation.[185] There is protection for 'whistle blowers' who have released environmental information.[186]

NEMA provides broad powers of entry and enforcement, including service of compliance notices.[187] In the interests of environmental justice, there is explicit provision for legal standing for individuals and groups, for review and private prosecutions, and that courts may decide not to award costs against such groups.[188] Such broad provision could usefully be made in other jurisdictions, though further specification might also be desirable. Where there has been a conviction, courts may order compensation and/or remedial measures,[189] and there is also outline provision for financial penalties instead of prosecution.[190]

---

[168] QWA Chapter 3, amended by Water and Other Legislation Amendment (Qld) Act 2010 No.53.

[169] QWA Chapter 3A, inserted by South East Queensland Water (Restructuring) and Other Legislation Amendment (Qld) Act 2012 No.39.

[170] Draft Technical Regulations for Petroleum Exploration and Exploitation General Notice 1032 of 2013.

[171] Eaton (2013).   [172] Constitution of South Africa s.44 and Schs.4, 5.

[173] Constitution of South Africa s.24.   [174] NEMA 1998 s.2.

[175] NEMA Amendment Act 2004 No.8.

[176] NEMA Amendment Act 2003 No.46.

[177] NEMA Environmental Laws Amendment Act 2008 No.44.

[178] NEMA (Waste Management) Act 2008 No.59 s.49, s.65.

[179] NEMA (Waste Management) Act 2008 Part 8.   [180] NEMA 1998 s.28.

[181] NEMA 1998 s.1.   [182] NWA s.22(3).   [183] NWA s.22(4).

[184] NEMA 1998 Chapter 3.   [185] NEMA 1998 Chapter 4.

[186] NEMA 1998 s.31; South Africa also has a general Freedom of Information law, Promotion of Access to Information Act 2000 Act No.2.

[187] NEMA 1998 ss.31A–Q.   [188] NEMA 1998 ss.32–33.

[189] NEMA 1998 s.3.   [190] NEMA 1998 s.34G.

South Africa does not have specialist courts, and usually environmental offences will be heard first in the Magistrates' Courts. In 2003/4 there was a brief experiment with a specialist environmental tribunal, focused on wildlife crime, but that did not continue, despite apparently good results.[191] In 2010/11 there was further consideration, but it was decided that the volume of case work did not justify the administrative burden; instead, there would be additional training for judges in the ordinary courts[192] – a similar approach to that in Scotland.

4.8.1   The National Water Act

Pollution is defined broadly under the NWA[193] to include direct or indirect alterations, which *inter alia* make the water resource less fit for beneficial use, or harmful or potentially harmful. 'Resource quality' is also broadly defined, to include water quality, flow, habitats and biota.[194] As discussed in Chapter 2, the National Water Resource Strategy (NWRS) must provide for water quality objectives, to be achieved by a classification system; Catchment Management Strategies must similarly take account of that classification and subsequent Resource Quality Objectives (RQOs).[195] The classification system and the RQOs are fundamental to the management of water quality in South Africa and will be examined further below.

There are two provisions specific to pollution, and then the general regime for controlling water uses will also apply. Persons in control of land, or occupying or using land, must take 'all reasonable measures' to prevent pollution. CMAs have powers of direction, and remedial powers, and may recover costs jointly and severally from anyone who 'was responsible for, or who directly or indirectly contributed to' the pollution, or the owner at the time or their successor in title, or any person in control of the land or anyone who negligently failed to prevent the pollution taking place.[196] Breach of a directive is a criminal offence;[197] appeals against directives and cost recovery are made to the Water Tribunal.[198] Where an emergency incident occurs, 'responsible persons' must report the incident to an appropriate authority and then take all reasonable measures to address the problem; again the CMA may issue directives or do works and recover costs.[199]

Water use under the NWA includes *inter alia* the disposal of waste and discharge of waste and wastewater into a water resource.[200] The general power to make regulations includes the management of waste, and protection for the resource and habitats.[201] The general licensing regime has been examined in Chapter 3, but pertinent to this chapter, licensing authorities must consider *inter alia* the class, RQOs and needs of the Reserve;[202] conditions may include *inter alia* emission levels, effluent treatment and monitoring.[203] Although the system for compulsory relicensing is focused on water quantity, quality is also a potential justification for this process.[204] There are specific Ministerial powers for 'controlled activities' including wastewater irrigation and aquifer recharge.[205] The pricing strategy for water uses may differentiate on the grounds of the class or RQOs, and may set different charges for different wastes.[206]

Powers of entry, for inspection and authorisation, are granted to authorised persons.[207] National monitoring systems must collect *inter alia* information relating to assessment of water quality, compliance with RQOs and the health of aquatic ecosystems.[208] There is a series of offences, including 'unlawfully and intentionally or negligently commit[ting] any act or omission' which pollutes or detrimentally affects a resource or is likely so to do. Unlike the corresponding principal offences in the other jurisdictions, this is not a strict liability offence, but the procedural offences, and the offence of using water 'otherwise than as permitted under this Act', are undoubtedly of that character.[209] The court has powers to award damages, order remedial works and inquire into the costs of making good any harm or loss.[210]

A major contributor to poor water quality and degraded ecological status in South Africa is the mining industry, especially acid mine drainage. The consequences of mining activities for water quality are significant enough to deserve some brief attention; acid mine drainage is only one of these.[211] Mining is a separate sectoral regime; in South Africa mining companies also require a water use licence under the NWA. Though acid mine drainage is also a feature of working mines, it is particularly problematic when emanating from closed facilities, where there may be additional issues of historic legal liability. Many mines in South Africa are no longer active, and modern regimes for closure and remediation may not have been in place, or the owners may no longer exist. An experts' report commissioned for DWA analysed the problem, and set out recommendations in

---

[191] AFROL News (2004) 'South Africa Sets Up New Environmental Court' (online) 24 February.
[192] OECD (2013) Chapter 2 section 4.6.   [193] NWA s.1.
[194] NWA s.1(xix).   [195] NWA s.6, s.9.
[196] NWA s.19; and see below on recent case law upholding these powers against a prior owner in the Supreme Court.
[197] NWA s.151.   [198] NWA s.148.
[199] NWA s.20. When a CMA is not established, or not yet exercising all its powers, these are exercisable by the Minister.

[200] NWA s.21.
[201] NWA s.26. Waste is defined in s.1 to include solid waste and water containing waste, and sediment.
[202] NWA s.27.   [203] NWA s.29.   [204] NW s.43; and see Chapter 3.
[205] NWA ss.37–38.   [206] NWA s.56.   [207] NWA s.125.
[208] NWA s.137.   [209] NWA s.151.   [210] NWA ss.152–153.
[211] Mining regulation, including closed or abandoned mines, is regulated by the Department of Mineral Resources under the Mineral and Petroleum Resources Development Act 2002 No.28.

the short and medium term.[212] These included pumping, preventive measures for the future, and a variety of active and passive treatment processes, all of which incur significant expense. A further report has been produced, but at the time of writing had not yet been published.[213] In a recent Supreme Court case, a directive by DWA under s.19 of the NWA to take remedial steps, issued to a mining company that no longer owned the land, was upheld as being within the scope of the Act and wider NEMA principles.[214] This is to be welcomed, but will not address the situation where the company causing the pollution no longer exists.

4.8.2 General authorisations

The relevant general authorisation permits irrigation with wastewater, discharges of waste and wastewater, and the storage and disposal of wastewater and greywater.[215] There are some general requirements that apply to all such uses, including having lawful use or access to the property; avoiding harm to others' property or water use, and any health and safety impacts for the public; requirements to monitor, sample and report on the activities; and 'precautionary practices' such as avoiding nuisance or contamination, and appropriate disposal of solids or sludge. Some uses above a certain limit must be separately registered before the authorisation applies.

Within each category of use there are volume limits as to how much wastewater can be stored, discharged etc., and these are linked to location and the strength of the effluent. Generally the rules apply to domestic wastewater and to biodegradable industrial effluents, e.g., from food processing. Irrigation, storage and disposal are not permitted within 50 m of the 1 in 100 year flood line or the riparian zone; 100 m from a watercourse; 500 m from a borehole or wetland; or above a major aquifer.

Discharges are permitted using general effluent standards for most waters and stricter special effluent standards for listed water resources. These general and special effluent standards are therefore a basic set of emission limit values, covering core physico-chemical parameters such as salinity, acidity and oxygen demand, along with faecal coliforms, nutrients and a short list of metals of particular concern. The WHO recommends the establishment of sets of basic emission and/or quality standards in developing countries, applying to parameters of particular local concern and, especially, physico-chemical and microbiological pollutants;[216] these should always be a starting point when developing pollution control regimes.

Disposal of wastewater is not permitted in specified control areas for groundwater. Disposal and storage are relevant to septic tanks, soakaways, pit latrines and other on-site sewerage; here, if the volume or density is above a certain limit, the local authority for the area must register the use. Finally, there is a new category of use in the 2013 rules that allows the removal of 'water found underground' – not groundwater as such, but e.g., water in basements or tunnels.[217]

As with the GBRs in Scotland, the exempt uses in Schedule 1 have less application to discharges, but will apply to waste or wastewater, or surface runoff, into 'conduits or outfalls' managed by others, who are in turn responsible for treatment and management. In Scotland there is a broadly similar three-tier structure, but whilst South Africa provides detailed rules for the general authorisations in the middle tier, in Scotland the only statutory specification is in the lowest tier. This is likely to change under the current regulatory reform, where more use will be made of standard rules, as is happening in different forms in all the jurisdictions.

4.8.3 The classification system and Resource Quality Objectives

The NWA sets out a framework for classification of water resources, and the setting of RQOs dependent on that classification.[218] This process feeds into the establishment of the Reserve,[219] because the ecological flow needs and water quality requirements will be higher for water of the highest classification. Together, these three elements are referred to in the surrounding policy context as 'resource directed measures', and work with the 'source control measures' such as licensing and technical standards outlined above. Perhaps unsurprisingly, the process has been much slower than originally anticipated; completing the process for every significant water resource is a key strategic aim under the NWRS2.[220] The policy environment has been developing over the last 15 years, with detailed early work on ecological classifications for rivers, wetlands, estuaries and groundwater,[221] and subsequent refinements.[222] For surface waters, there will be an ecological assessment of water bodies and a six-fold categorisation: A, unmodified natural; B, largely natural; C, moderately modified; D, largely modified; E, seriously modified; F, critically modified. Categories E and F are considered unacceptable and should therefore be remediated. Categories A–D will map onto three management classes, I–III, and these have now been defined in regulations.[223] Class

---

212 Coetzee *et al.* (2010).   213 DWA (2013b).

214 *Harmony Gold Mining Company Ltd v Regional Director: Free State Department of Water Affairs* (971/12) [2013] ZASCA 206.

215 General Authorisation No.665 of 2013.

216 Helmer and Hespanhol (1997).

217 General Authorisation No.665 of 2013.   218 NWA ss.12–15.

219 NWA ss.16–20.   220 NWRS2 section 5.4.1.

221 See for the early work, DWAF (1999).

222 See DWAF (2007); Kleynhans and Louw (2007).

223 Regulations for the Establishment of a Water Resource Classification System 2010 No.810.

I is 'minimally used', and minimally altered from its pre-development condition; Class II is 'moderately used' and moderately altered; and Class III is 'heavily used' and altered. The management classes are wider than the ecological assessment and the process will include all the socio-economic factors that are relevant to the allocation of water and approval of water uses under the NWA.[224] These are the classes that are part of the legal process outlined above and will be used to determine the specific RQOs for each water body, and the ecological Reserve. The system in South Africa therefore has more similarity to Queensland than to the EU; there is no general presumption of achieving a particular class beyond remediating the most damaged systems. Instead there is a recognition that ecological class will be traded off against socio-economic uses, and RQOs set accordingly. Such an approach may be more honest than the EU system, which sets an overall aim and then provides for exceptions.,

Also similar to Australia, there is a pre-existing set of water quality guidelines, covering a very broad set of parameters.[225] The freshwater uses covered are domestic use, aquaculture, irrigation, livestock, recreational use, industry and aquatic ecosystems. These include guidelines for physico-chemical and chemical substances, and although now a little out of date, these will form part of the assessments for determining ecological class.

4.8.4 Groundwater

Although groundwater is not easily subjected to ecological classification *per se*, early work set out methodologies for classification and establishing RQOs.[226] Relevant factors for classification were identified as being the volumes of abstraction, levels of contamination and the impacts of various land uses, with potential RQOs including no decline in level or quality, and no impact on surface flows. This has significant correlation to the approaches to groundwater management under the WFD. The surrounding policy context includes a recent strategy, which sets out the importance of groundwater to South Africa, especially for drinking water; the threats to its sustainable use; the legislative framework; and also specifies the volume of available resource, nationally and for the 19 current water management areas.[227] There is also a technical guideline for those actually withdrawing and using groundwater, including private industry and water service providers, which contains guidance both on management, such as participation and the linkages to the wider IWRM process, and on technical matters such as drilling boreholes.[228]

4.8.5 Diffuse pollution

In South Africa, both salinity and nutrient runoff are recognised problems for water quality by DWA.[229] The Department of Agriculture, Forestry and Fisheries is responsible for farming regulation and policy, and there is a LandCare programme, and a programme for Natural Resources Management, drawing on the Australian model.[230] There is some Government lending to emerging and other poor farmers at low cost through a Land Bank, but without ties to environmental improvements. Unsurprisingly, the Department's focus is on supporting emerging black farmers and there is perhaps less emphasis on the management of fertilisers and pesticides than in the other jurisdictions. The Department issues guidance on agricultural practice, but the main concern is to enable farmers to meet hygiene etc. regulations, in order to enter markets at home and abroad.

The major science research centres provide analysis of water quality problems including diffuse pollution and agricultural impacts, including both the Water Research Commission and the Agricultural Research Council, public bodies established by statute.

4.8.6 Water quality and current initiatives in ecological protection

The ecological approach is evolving in South Africa as in other countries. There are many initiatives, some under the control of the DWA and specific to water, others relevant to NEMA and primarily managed by the DEA, especially in relation to biodiversity. The DEA runs a suite of programmes that combine working and training with environmental protection, especially 'Working for Water' (previously managed by DWAF) and 'Working for Wetlands'.[231] Finally, recent collaborative work by several departments, agencies and NGOs has resulted in the identification of a set of National Freshwater Ecosystem Priority Areas.[232] These are freshwater systems (rivers, wetlands, estuaries) that merit special protection to ensure the survival of South Africa's biodiversity, and some unimpacted systems, despite the multiple pressures on water quality and water use. Thus South Africa has a structure and systems for the management of water pollution and water quality that reflect best practices in other countries with more advantages; but in every jurisdiction the problems outpace the solutions.

---

[229] 'WQM in SA' see generally http://www.dwaf.gov.za/Dir_WQM/wqmFrame.htm.

[230] 'LandCare' and 'Natural Resource Management' Programmes see generally http://www.daff.gov.za/daffweb3/Programmes.

[231] 'Working for Water' and 'Working for Wetlands' Programmes see generally https://www.environment.gov.za/projectsprogrammes.

[232] Driver *et al.* (2011); the bodies include the South African National Biodiversity Institute, the Council for Scientific and Industrial Research, the Water Research Commission, the World Wide Fund for Nature, the South African Institute for Aquatic Biodiversity and the South African National Parks, as well as DWA and DEA.

---

[224] DWAF (2007b).   [225] DWAF (1996).   [226] DWAF (1999a).
[227] DWA (2010).   [228] DWAF (2008a).

4.9 ENFORCEMENT AND SANCTIONS

It seems appropriate to end this chapter with a brief note on actual enforcement measures. In South Africa, DWA monitor and investigate water-related activities, with especial focus on wastewater treatment plant and mines.[233] Statistics on enforcement actions as such for all media are given in the DEA's annual report: in 2012/13, 81 incidents were notified and 72 investigated; 81 facilities inspected; 69 enforcement notices were issued with 82% compliance; and 21 criminal investigations were reported to the prosecutor.[234] Given the population size the figures seem small, but at least they do report these. In the past, SEPA's annual report contained comparable data for each control regime, but this is no longer done; they do report separately on enforcement, but the last such published report was 2011/12.[235] In that year, there were 37 reports to the prosecutor, 124 statutory notices and 160 'final warning letters', across all the control regimes (mainly water, waste and industrial air pollution). Typically in Scotland, most prosecutions for water offences are against SW, which holds by far the greatest number of discharge consents. In Queensland, the DEHP report annually in some detail. In 2012/13, they issued 168 penalty notices, 51 'transitional environmental programs'[236] and 28 environmental protection orders, again over all the media covered in the EPA. There were 11 prosecutions. The majority of penalty notices were for contravention of conditions; one of the prosecutions was for breach of development consent relating to wastewater treatment plant.[237] In England, there were 407 prosecutions of which 55 related to water, 204 cautions with 61 for water, and 102 notices, with 8 for water. There is no systematic reporting of enforcement data in England.[238] The high number of prosecutions may reflect the fact that EA staff can bring prosecutions in the lower courts. In Queensland, there is a greater variety of penalties available across different control regimes and these seem to be used in preference to prosecution, which would be expected where the regulator has effective sanctions in their own hands. If civil sanctions are extended in England, or introduced in Scotland, they would be a useful addition.

4.10 CONCLUSIONS

What conclusions can be drawn in relation to water pollution and water quality? All the jurisdictions have broadly comparable regimes for environmental protection. In Queensland, the

pollution regime is the most highly integrated and there is a coherent system for development consents and environmental licensing. The overarching policy concept of ecologically sustainable development has a statutory basis. In South Africa, the constitutional provision is radical, and the hierarchy of Constitution, NEMA and NWA is highly structured, again with express high level principles. The operational regime for water predated other environmental reforms. It is therefore possible to reform all aspects of water law, including pollution control, even in the absence of a review of wider environmental law, and this approach can enable priority for the water agenda. In England, the integrated regulations provide a single control regime for discharges to water, and there is *de facto* integration through the regulator. There is a proliferation of other specific controls as well, in part due to the structures for implementation of EU law, but this is less so with the implementation of the WFD and the repeal and/or integration of some other EU rules. EU law has provided an important driver for regulating all aspects of water quality in the UK. In Scotland, there is much more integration now, under WEWS, across the operational areas of water management. Again, this is evidence that it is possible to take forward a water law reform agenda in the absence of concomitant reform in other areas. Even in Scotland there remain a large number of subsidiary rules that are not fully integrated into the general regime. In the UK jurisdictions, the independent regulator provides cohesion, consolidates resources, and can perhaps be a distinct 'voice' in the policy arena that is lacking where regulators are government departments. In addition, the latter model will inevitably be spread across departments, and on these two grounds a separate regulator may be an advantage.

Whether there is an integrated environmental regime or an integrated water regime, but especially the latter, it will always be necessary to provide for inter-agency communication. Controls on waste disposal have consequences for water, in particular groundwater, but so too do rules on contaminated land, industrial air pollution and various sectoral uses of water; the same is true for legal and policy frameworks for biodiversity or climate change. Queensland probably has the most specificity in this regard; in South Africa, NEMA provides high level strategic mechanisms.

All areas make use of a combination of environmental standards, with particular emphasis on emission limits and quality standards. The former are simpler to develop and apply, although still necessitate regulatory resources, and should be the starting point for site-specific controls. Quality standards should be set firmly in the framework of water resource management and the concomitant assessment of the resource base. In both Queensland and South Africa, these quality measures take the form of guidelines. The historic experience in the UK jurisdictions suggests that mandatory standards are preferable, otherwise there is scope for regulatory flexibility and, consequently, for pressure by

---

[233] DWA (2013c).   [234] DEA (2013).   [235] SEPA (2012a).

[236] These provide a binding programme for an operator to achieve phased compliance; EP Act Chapter 7 Part 3.

[237] DEHP (2013a).

[238] These statistics were obtained by FOI request, EA (2014).

operators on regulators. In Queensland, the design allows variation to set more specific objectives through local water resource planning mechanisms, with the safeguard that there should always be an improvement in water quality; this may be a good compromise.

Although quality standards are more complex and expensive to develop and monitor, they provide more comprehensive analysis of water quality and assist with the management of diffuse pollution. Where resources are particularly restricted, states should focus on providing for physico-chemical standards (oxygen demand, alkalinity, temperature etc.) and quality standards for a limited range of chemical pollutants that are most prevalent, and/or most affect the basic uses of abstracted water, in a given location. As new pollutants emerge, and more evidence appears as to their effects in combination and over time, quality standards may need to be introduced or tightened, but this has financial consequences either for industry directly or for urban wastewater management (especially for managing impacts from pharmaceuticals and personal care products). In South Africa and Queensland there is a single set of quality guidelines, which seems a more straightforward way of organising these controls than the multiple instruments used in the EU.

Regulatory resources will always be under pressure, especially but not exclusively in developing countries. Licensing regimes are expensive to administer and, even where these are notionally self-funding, high charges for users are likely to be equally problematic. The use of general rules or binding codes, which can be activity- rather than operator-specific, is increasing in all the jurisdictions. These should be mandatory and backed up by appropriate criminal sanctions, so it will not be the case that regulation is not required. This may allow regulators to divert more resources to education and, where necessary, enforcement; but its success is also heavily dependent on adequate monitoring. The rigour of a full licensing regime should be reserved for the largest, most complex or potentially hazardous processes and activities. This may give a better 'spread' of regulatory effort and maximise the effectiveness of resources. That should also enable maximum transparency in the issuing of full licences, for example in public consultation. Specific pollution control issues from the extractive industries, both new and emerging, require a coherent legislative framework, but also increase tensions between environment and development.

Agriculture is a significant user of both land and water and (with other rural land uses) is a major contributor to diffuse pollution. If relevant EQS have been established, these will apply to water affected by farming as with other activities. The use of general rules or mandatory codes is supported, with a clear strategy and policy from the legislature to assist in changing behaviour. There is a difference between initiatives for behavioural change in a voluntary world, and those used alongside mandatory codes. In the latter situation, there are sanctions and

the basis for regulatory involvement is clearly established; this enables proper enforcement where there are persistent breaches. The benefit of legislation may be as much that it mandates regulators to act. Where there is subsidy assistance to farmers, this should certainly be tied into the achievement of environmental objectives.

In all the jurisdictions there are broadly comparable powers of entry and enforcement, with growing interest in financial penalties directly applicable by regulators, rather than criminal penalties in the courts. All have rules on access to information, and some provision for consultation on licence applications. These legal processes and structures are well established, and generally effective in tackling point source pollution, assuming sufficient resources are granted to the regulators.

Queensland is the only jurisdiction with a specialist court with general authority in most environmental matters, though England has a new tribunal. Specialist courts may have a better appreciation of the environmental issues and ideally would have wide jurisdiction over less serious criminal prosecutions, regulatory appeals and first stage review of regulatory decisions, giving a comprehensive decision-making body. Perhaps more important are an appropriate range of penalties that will provide a real deterrent, and clear statutory rules on (affordable) access to justice and judicial review. Given the complexity of modern environmental law, a state-funded environmental law agency must be a positive recommendation, and it is sad that Australia is withdrawing funding for this.

All of the jurisdictions are introducing systems for ecological classification, in the context of reform of water resource management and linked to quality standards for specified uses. In all the regimes, the broad approach is the same, with undisturbed water bodies used as reference conditions to determine appropriate standards of ecosystem health, with standards set for chemical toxicants, flow, and interference with the structure of the river, to maximise ecosystem potential. In the EU, there is an overall objective of 'good' quality along with exemptions and extensions for modified waters as a result of human activities. In South Africa and Queensland, the classification standard to be achieved depends on the use of the waters; this achieves the similar result, of accommodating the need for sustainable human developments, and may be more straightforward. In all cases it is accepted that the desired improvements will only be achieved through wider resource management processes. Ecological classification is demanding in human and other resources and is a progressive achievement, certainly in developing countries, but there is a synergy here; the assessment and monitoring required by the introduction of IWRM will provide some of the information required to begin this process.

In Queensland and South Africa, although the water quality guidelines provide for basic physico-chemical standards as well as chemical toxicants, they do not address the level of flow, or the measures for the protection of habitats and biota. In Queensland, these are addressed across the two sets of resource plans set out in Chapter 2, and also through the specific objectives set under the Policy, which may use the measures in those resource plans. In South Africa, the RQOs will address all these factors once established. In the UK and throughout the EU, the WFD and the river basin plans will bring together the ecological standards into the programme of measures, along with the application of the chemical (priority substance) standards and any special requirements in other Directives for protected areas. An ecological approach brings together all the elements of this book, but the environment remains a wicked problem, with complex externalities poorly managed. The promulgation of new law is rapid, but can only be one tool. If there is no social or political consensus around the relationship between environment and development, then there will be no solutions. The pressures (population, climate, urbanisation) continue to increase; enshrining sustainable development in law has not yet achieved it as a policy outcome.

As we move on to consider the provision of water services, both quantity and quality are critical factors. Water services providers will abstract water under the rules in Chapter 3, but these services are dependent not just on an adequate supply of water, but on water of an appropriate quality; the WFD and the Australian water quality guidelines both make specific reference to protecting abstraction sources to require minimal subsequent treatment. Water safety planning and catchment protection are part of water quality management and of service provision, bringing us back again to the fundamental need for planned management of the resource in all its uses, and highlighting again the interdependency of the core operational areas.

Table 4.1 *Key findings Chapter 4: water pollution and water quality*

	South Africa	Queensland	England	Scotland
Regulator	Department	Department	Agency	Agency
Integrated environmental regimes	Highly integrated under NEMA, but not for water consents.	Highly integrated under EPA; all media and with development consents.	Integrated under EPR; and through single regulator.	*De facto* integration through single regulator (moving to integrated permits).
High level principles/ duties	Human needs; SD; equity; justice; integrated environmental management (NEMA). (WRM duties Table 2.1.)	Ecologically sustainable development (EPA and QWA). (WRM duties Table 2.1.)	SD; pollution control; conservation/management of water resources (Environment Act).	SD; pollution control; (Environment Act). (WRM duties Table 2.1.)
Integrated water use licences	Yes	No	No	Yes
Courts and tribunals	Ordinary courts.	Planning and Environment Court.	First tier Tribunal; ordinary courts.	Ordinary courts; specialist prosecutors.
Standards	Emission standards, EQ guidelines (combined set).	Emission standards, EQ guidelines (combined set).	Emission standards, EQS (specific to water use).	Emission standards, EQS (specific to water use).
Ecological classifications	Developing	Yes	Yes	Yes
Diffuse pollution	Indirect controls.	Specific controls for reef; indirect controls and NRM planning.	Indirect controls; farming subsidies.	Specific controls under CAR; indirect controls and farming subsidies.

5 Governance and regulation of water services

5.1 INTRODUCTION

This final substantive chapter will investigate the remaining operational area of a national water law as defined, water services. In this book, this term is used to include the supply of drinking water and wastewater services, along with supply of piped water to industry (urban water services), and provision of basic sanitation. The policy context discussed in Chapter 1 frames the provision of the service as well as the management of the resource, and in many ways the crisis around service provision has driven these global agendas and kept them prominent.

The UN currently estimates that out of a global population of some 7.2 billion, some 768 million people still lack access to an 'improved' water service, and 2.5 billion lack 'basic' sanitation.[1] The terminology is fluid. The original Millennium Development Goals (MDGs) in 2000 set a target of halving 'the proportion of people who are unable to reach or to afford safe drinking water'.[2] At the Johannesburg Summit, the international community added a commitment to halving the numbers without 'basic' sanitation.[3] The current formulation of the Goals refers to halving 'the proportion of the population without sustainable access to safe drinking water and basic sanitation'.[4] 'Improved' services have been defined,[5] but there may be an improved service for which there is insufficient access, for example where water is only available for short periods or when a long journey must be made to collect water; similarly, an improved supply that has not been maintained may not stay 'safe'. Disturbingly, some analysis suggests that the numbers without safe supply may be twice those without an improved supply.[6] Data and reporting are problematic and there are significant disparities within countries, between rich and poor, urban and rural, and for women, the sick and disabled, and minority ethnic groups.[7] The cross-cutting impacts and benefits of investment in water supply, sanitation and hygiene have been discussed in Chapter 1; the most recent synthesis report from UN-Water suggests that whilst the necessary investment to achieve universal basic coverage would be USD535 billion, every dollar spent on water supply leads to 2 dollars' benefit in health alone, and for sanitation, the return is 5.5 dollars.[8] Globally, neither the scale of the problem nor its importance can be underestimated.

Traditionally, provision of drinking water, as with other essential services, has been dominated by the state; the provider has often been local government, or some sort of agency or appointed board, typically regulated by a central government department. In recent years, there has been much interest in attracting private sector involvement, and this can be achieved in different ways. This chapter will use the general term water services provider (WSP) to apply to all providers of these services; where there is separation between bulk supply and retail distribution, WSP will be used to refer to the retail provider. Where specific legislation uses specific terms, those will be used in context.

A principal reason for seeking private sector involvement is funding. The water services industry is highly capital-intensive and the lifetime of projects such as treatment plant may be anywhere between 25 and 40 years. Cost recovery is made more difficult by the political imperative to provide water services, which may have been highly subsidised in the past, leading to resistance to increased charges to meet investment; but without investment, networks cannot be maintained, or expanded to areas without supply. Significant work was done a decade ago to identify the amount of investment required, especially to meet the needs of the unserved poor.[9] Certain key themes emerge, including the need to make any cross-subsidies transparent, for

---

[1] WHO/UNICEF (2012); UN-Water (2014). [2] UN (2000) para.19.
[3] UN (2002) para.25.
[4] 'Millennium Development Goal 7'; see http://www.un.org/millenniumgoals/environ.shtml.
[5] UN-Water (2006). As regards water supply, the following would be improved: piped water, water from public standpipes, protected wells and springs, rainwater and bottled water (although the latter is generally too expensive for the urban poor). Unimproved sources would be unprotected wells and springs, water from vendors and tankers, and from surface waters. For sanitation, improved provision is flush or pour-flush toilets to sewer, septic tanks, ventilated pit latrines, pit latrines with slab, and composting toilets. Unimproved sanitation is public or shared latrines, pit latrines without slabs, hanging toilets or latrines, buckets, or no facilities at all.

[6] UN General Assembly (2013) A/HRC/24/44.
[7] UN-Water (2014); WHO/UNICEF (2012); WHO/UN-Water (2012).
[8] UN-Water (2014).
[9] See, in particular, Kessides (2004), Winpenny (2003), PPIAF (2001).

example between rural and urban users, or industrial and domestic users. Another is a preference for subsidising connections rather than reducing tariffs as such, though stepped tariffs may be useful (for example a free or low cost supply of a minimum amount of water) and two-part tariffs, with a separate element for infrastructure and volume supplied, may assist with transparency of cost recovery. Most relevant to this book is the emphasis on effective regulation, regardless of whether providers are in the public or private sectors or a mixture of both.

There is an intense and highly polarised political debate over private sector involvement in water services. Some critics would say that the World Bank promoted 'privatisation' on ideological grounds, as part of the so-called Washington consensus (or neoliberal economic agenda) driven by the British and American Governments in the 1980s and 1990s.[10] Undoubtedly there was a strong ideological drive for the involvement of the private sector at that time, not just in natural monopolies, at least in the UK;[11] and the Bank's approach did reflect that,[12] as did the fourth Dublin Principle, that water is an economic good.[13] Lending for large projects became increasingly conditional on achieving policy reform, including private sector engagement and full cost recovery for services;[14] this further polarised debate.

The private sector can be involved in service provision in different ways, with varying levels of risk, responsibility and potential profit. At the lower end, there is widespread use of short-term service contracts, perhaps to supply equipment or vehicles, or to perform specified operational or maintenance tasks. At the other end of the scale, major concessions transfer all responsibility for the whole system, including investment risk, for an extended period, perhaps 30 years, but do not transfer ownership of the assets. A lease, or *affermage* under the French model, may transfer the whole operation but without investment risk. Also important are contractual mechanisms such as build–own–operate (BOO), build–own–operate–transfer (BOOT) and similar variants, sometimes also described as concessions, which are usually for treatment plant (water or wastewater), likely to last for 20–25 years, and these may share risk under the contract in various ways. Finally, complete divestiture of the asset base is rare, but has been done in England.[15] The term 'privatisation' can be used to mean divestiture, but is also used to mean any and all of the many contractual variations on these basic 'models'. For that reason, the chapter will use private sector participation (PSP) as a general term, to cover all engagement, and public–

private partnership (PPP) where there is a contractual arrangement, including BOOT-type schemes but also, for example, joint ventures. Usually, we think of PSP in networked services as involving major international players, but in addition there is a role for smaller-scale local providers, and this may be less contentious than transnational investors.[16] The chapter will use 'divestiture', but will avoid 'privatisation' except in the context of that political and polarised debate.

Recently, there is evidence of a shift in the type of PSP taking place in the water sector. Whilst all international, and national, investment in infrastructure has suffered as a result of the global financial crash, changes in investments in water were already happening, prompted by high profile failures of contracts in Argentina, Bolivia and Tanzania.[17] The result has been much less interest in large-scale whole system concessions, with transnational companies preferring either BOOT-type schemes or management contracts; the latter provide advice and support, are usually short term, may have performance payments, but do not take responsibility for the system.[18]

BOO and BOOT schemes have several advantages for the investor, especially given the political context. They usually relate to a single plant, and are essentially construction contracts, which may have an operational element. The contract is usually with the public authority that has responsibility for providing the service, so there is no need to recover charges directly from the public, and indeed the public may not be aware that the contract exists or the plant is in private hands. The networks, which are least likely to be profitable, are the part of the system least susceptible to PSP.

Given the contentious policy environment, it is perhaps helpful to remember that the overall share of the private sector is very small; Marin estimated that it was around 7% in 2007. Even if it is a little higher, to account for smaller service contracts that he did not consider, 15% would be a generous estimate. Although the types of contracts have changed, divestiture remains a very unlikely option. Overall, most assets, and most responsibility for services, remain in public hands.

One of the effects of two decades of focus on PSP was an interest in economic regulation. As water is a natural monopoly, and an essential service, there will be little competitive pressure to keep prices and profits down. One response is to introduce competition to those parts of the sector that are susceptible to such pressure. Another is to regulate the provider and control

---

[10] See, e.g., Finger and Allouche (2002), especially Chapter 3. For a general critique of private sector participation, see, e.g., the work of the Public Sector International Research Unit at the University of Greenwich, especially Hall and Lobina; see www.psiru.org. Specific to the EU's role, see World Development Movement (2007).

[11] Vickers and Yarrow (1988).   [12] World Bank (1993).

[13] Dublin Statement (1992), and see Chapter 1.

[14] Olleta in Cullet *et al*. (2010).   [15] See Delmon (2000, 2001).

[16] Malacek (2013).

[17] *Compañía de Aguas del Aconquija S.A. and Vivendi Universal S.A. v Argentine Republic* (ICSID Case No.ARB/97/3); *Biwater Gauff (Tanzania) Limited v United Republic of Tanzania* (ICSID Case No.ARB/05/22); *Aguas del Tunari S.A. v Republic of Bolivia* (ICSID Case No. ARB/02/3); and see Solanes and Jouravlev (2007).

[18] Marin (2009).

either the price, or (usually) the return on capital.[19] It is also possible to apply principles of economic regulation to the public sector. The inefficiency of the public sector was ostensibly the reason for the great free-market crusade of the 1980s, yet there is evidence that the public sector, properly regulated, is not necessarily less efficient;[20] this chapter will take as a starting point the presumption that regulation is more important than ownership. In England, the service is divested; in Scotland, South Africa and Queensland, there is a public service with PSP in various forms, and some commercialisation or corporatisation, whereby commercial principles are applied to the public providers.[21]

The water services industry is regulated in terms of drinking water quality and other service standards, environmental controls for both abstractions and discharges, and the economic regulation of tariffs and investment programmes. Thus their use of water should be planned within the IWRM processes set out in Chapter 2, whilst environmental protection is achieved by the control of abstractions discussed in Chapter 3 and the regulation of discharges in Chapter 4. A narrow view of economic regulation would require that charges cover only the costs of delivering the service standards, maintaining the system, and allowing some return on capital to ensure investment in the future. A broader view of economic regulation would include the environmental costs, and also some social protection for the poor and unserved. This chapter will presume that wider approach, to internalise environmental externalities and help to realise the goals of the global policy debate.

The mode of regulation will depend in part on the models chosen for service delivery and on the wider political, social and economic context. Regulatory provision may vary in structure – water regulator, general utilities regulator, environmental regulator, competition commission and government department(s) are all possibilities, and indeed several of these may have different roles. An independent agency may give a degree of separation from political influence, although where the WSP is in the public sector, the government may want the final say over prices. In a global context, the World Bank has proposed some safeguards for (relatively) independent regulators.[22]

Just as effective regulation is necessary for both public and private sector delivery, so too is good governance. Governance was discussed in relation to water resources in Chapter 2, but is highly applicable to service delivery. In this context, the term may refer both to broad political mechanisms relevant to the control and management of regulated industries,[23] and to the specific attributes of transparency, participation and accountability in decision-making. These will be central to the analysis in this chapter. Linked to an absence of both regulation and governance, corruption is a problem in water services as in other capital-intensive sectors, and although it will not be possible to explore the topic in its own right, both regulation and governance help to address this at every level of the supply chain and of political activity.[24]

One final introductory issue is that of the human right to water; this has become a major source of debate.[25] Although there is no specific human rights treaty addressing water, or sanitation, the right to water is explicit in two of the core human rights conventions, one of which also includes sanitation, and is implied into others.[26] Following a General Comment in 2002,[27] the UN appointed an independent expert, now special rapporteur, and she has issued a series of reports, including on the right to basic sanitation and the role of the private sector.[28] The General Assembly has since passed several resolutions in support of the right, and so has the Human Rights Council.[29] So, absent its own treaty, it is becoming accepted that there is a customary international law right to water, though sanitation may not quite be at that stage. The UK Government has been particularly resistant, along with the USA, Canada and Australia.

The debate around the human right to water has several implications. For water law and water lawyers, there is a potential tension. If the human right is to be given priority, that may cast some doubt on the use of the IWRM process to allocate water to the most appropriate uses.[30] Domestic supply is a small proportion and is usually prioritised anyway, de facto or de juris, but, if the right extends to small-scale subsistence use as some argue, then the proportion thus allocated is much higher.[31] Also, where there is a strong constitutional regime for protection of property rights, for example in the USA, or within the European Convention on Human Rights (Chapter 3), then the arguments

---

[19] Ogus (1994) Chapter 14.

[20] Renzetti and Dupont 'Ownership and Performance of Water Utilities' in Chenoweth and Bird (2006).

[21] This is not without its critics; for a critique focused specifically on South and Southern Africa, see McDonald and Ruiters (2005).

[22] World Bank/PPIAF (2006); this is less prescriptive, and makes fewer assumptions about the desirability or feasibility of independent regulators, than the previous version.

[23] See, e.g., Stern and Holder (1999).

[24] Transparency International (2008).

[25] See, e.g., McCaffrey (1992), Winkler (2012).

[26] So the Convention on Elimination of Discrimination against Women (1979) specifies rights to water supply and sanitation; the Convention on the Rights of the Child (1989) specifies rights to clean drinking water. The Convention on Economic, Social and Cultural Rights (1966) (ICESCR) specifies rights to an acceptable standard of living, food and health, into which water can be implied.

[27] UN Committee on Economic, Social and Cultural Rights (2002) General Comment No.15 (GC15) on the Right to Water; GC15 stated that the right to water was implied into ICESCR, which as one of the general human rights treaties is of great importance.

[28] UN General Assembly (2009); UN General Assembly (2010).

[29] See, e.g., UN General Assembly Resolution (2010), UN HRC Resolution (2010).

[30] Tremblay (2011).

[31] Winkler (2012) argues for this, as does GC15 para.7, linked to the right to food.

may be used to protect existing commercial abstraction rights in developed countries, which is a very different debate that should not be conflated. It is useful to note the specific work of the special rapporteur on PSP, where she makes it clear that, where this occurs, the role of the state is to ensure that the right is still delivered by the private provider.[32] In other words, its regulatory role will always remain however the service is delivered.

In terms of access to water services, especially drinking water but also basic sanitation, when the South African constitution was drafted in the 1990s it was very unusual to find a constitutional right to water, or to a clean environment. Now these rights are much more common, albeit usually qualified and progressive. In this book the human rights debate is most relevant in South Africa, and will be explored below.

This chapter will examine the legislative provision for water services focusing on the primary and secondary legislation and policy documentation. It sets out a structure for a reformed water services law, identifying essential provision that states will need to make, and comparing possible approaches. These essential elements are: structure, ownership and control; duties of supply and service standards; economic regulation and business planning; and water conservation.

5.2 STRUCTURE, OWNERSHIP AND REGULATION

This section will consider the structure and ownership, and outline the regulation, of water services. In all the jurisdictions except England, water services remain in the public sector, with some variable degree of PSP. In South Africa and Queensland, there is vertical disaggregation between bulk supply and retail and distribution to the end consumers. The latter are provided primarily by local government but also by other bodies, of various types. In Scotland, there is a national monopoly with minimal liberalisation and some PPP in wastewater treatment; in England, there are regional monopolies with competition by comparison and some direct competition being introduced. In every jurisdiction there are powers to issue directions and take enforcement action by government and, in the UK especially, by other regulators.

5.2.1 England

In England, there was full divestiture of the asset base in 1989,[33] as part of a general process of 'privatisation' of public services and industries in the 1980s, although government retains regulatory control and water companies are licensed to supply services, licences which could theoretically be revoked. The industry is vertically integrated but with regional monopolies, and there is a separate economic regulator (now, the Water Services Regulation Authority (the Authority), but still generally referred to, by itself and others, as OFWAT).[34] Ministers set the policy context, the EA controls abstractions and discharges, and there is a separate Drinking Water Inspectorate.

There are currently 10 water and sewerage undertakers, a number of smaller water only companies and a number of new licensees. Most of the discussion in this chapter will refer to the undertakers, and will not refer to Wales. Although the legislative framework is broadly the same, and the Welsh incumbent has a private corporate structure, it now operates as a company limited by guarantee, with a different ethos.[35] Only around 1% of the population have private water supplies in England and Wales,[36] and around 2% have private wastewater treatment, mainly septic tanks;[37] this reflects the high degree of urbanisation and high population density.

In 1991 there was consolidating legislation in the shape of the Water Industry Act (WIA)[38] and WIA as amended remains the principal legislation. Major amendments have come in the WIA 1999, the Water Act 2003 (WA2003) and the Floods and Water Act 2010. The Water Bill 2013, before Parliament at the time of writing, will make extensive structural change and is being heavily criticised for its complexity and the failure to produce consolidating legislation.[39]

Full divestiture of water services is unusual globally, with most states preferring to retain long-term ownership of the asset base, but England has served as one model. Where ownership as well as management is divested, then regulation becomes the principal role of government. The companies are licensed to provide services, with licences lasting at least 25 years and in theory terminable by the state on at least 10 years notice, and with default provisions to ensure constancy of supply.[40] The conditions of appointment on which OFWAT appoints undertakers are effectively a parallel control regime operating in addition to the WIA.[41] Disputes over conditions of appointments are determined by the Competition and Markets Authority.[42]

---

34 WA2003 s.34 brought in a regulatory board to carry out the functions of the Director General of OFWAT, in whom powers had resided.
    A regulatory board is now considered in the UK to be better practice. This chapter will generally use OFWAT, except where discussing specific statutory provisions which use 'the Authority'.
35 'Dwr Cymru Welsh Water' see generally http://www.dwrcymru.com/en/Company-Information.aspx.
36 'Private Water Supplies' see generally http://dwi.defra.gov.uk/stakeholders/private-water-supplies/index.htm.
37 DEFRA (2003).   38 Water Industry Act 1991 c.56 (WIA).
39 Hansard HL [Vol.752] Col.106–109.
40 The special administration procedures under WIA 1991 ss.23–26.
41 'Instruments of Appointment' currently in force are available at https://www.ofwat.gov.uk/industrystructure/licences/; the power to impose conditions is in WIA s.11.
42 WIA ss.12–17.

---

32 UN General Assembly (2010).   33 Water Act 1989; Bakker (2003).

England enabled the introduction of common carriage, under the WA2003, for commercial users using more than 5 ML/annum.[43] This contrasts sharply with the view taken in Scotland, that the introduction of competitors' water into the mains presents a public health risk such that it should be prohibited. 'Inset appointments' were already provided for under the WIA and these enabled undertakers from different areas to supply individual customers in another undertaker's area.[44] The common carriage provisions extended this still further. 'Licensed suppliers' may be granted a retail licence to supply retail services, or a combined licence to provide retail services and insert a new water source into the undertaker's system.[45] Undertakers have a duty to provide access to the networks subject to certain qualifications[46] and OFWAT determines disputes over access and charges.[47] The regime came into operation at the end of 2005, and currently there are eight licensees.[48]

These attempts to introduce competition have been very problematic, especially the access arrangements. The legislation provides for the so-called 'costs principle',[49] by which incumbent operators are protected by being able to levy network charges minus only their avoidable costs – those they will not incur by serving a particular eligible customer. This leaves little margin for new entrants and has been heavily criticised in the courts in relation to the pre-existing inset regime. Indeed, the Competition Appeals Tribunal criticised both the incumbents and OFWAT, holding that the methodologies used to determine access prices were unsatisfactory, and the result amounted to the abuse of a dominant position.[50] The newest reforms in the Water Bill go further; they create separate retail and wholesale licences, for water or sewerage, and are intended to work with the introduction of upstream abstraction trading to further facilitate new entrants.[51] There are also earlier provisions (but not yet in force) to enable regulations to require large infrastructure projects, of a

scale that might affect an undertaker's ability to supply their customers, to be put out to tender.[52] Given the vertical integration and regional monopoly structure, this is an unusual provision.

A system of enforcement orders may be used, by OFWAT or the Secretary of State, to ensure either performance of a statutory duty or compliance with a condition of appointment.[53] Alternatively, the regulator may accept an undertaking from the company.[54] The WA2003 provides for financial penalties for contravention of conditions of appointment, contributing to another company contravening the same, or failure to achieve performance standards.[55]

There is a procedure for the making of special administration orders in the event that a WSP becomes insolvent.[56] These are made by the court on the application of the Secretary of State or OFWAT and will ensure that the statutory functions of the WSP are still carried out by another company.

The system in England is deceptively straightforward in terms of ownership, but the detail is complex and both regulation and governance have been problematic. Whilst most of this chapter is relevant to aspects of governance, especially accountability, there has been one specific issue in England around access to information. Until very recently, the water companies (supported by the Information Commissioner) have successfully argued that they are not 'public authorities' for the purposes of environmental disclosure.[57] The European Court of Justice has recently ruled on this, and though the decision is returned to the Tribunal, and is not particularly easy to read, it does tend towards the view that they are carrying out public functions and within the control of public authorities, and the rules should presumably apply.[58] More general concerns currently are about the financial structure of the undertakers, their gearing and level of returns to stakeholders, and these will be considered below.

5.2.2 Scotland

In Scotland, the public water supplier Scottish Water (SW) is a public corporation, vertically integrated and serving almost the whole population (over 95% of the population for water, and

---

[43] WIA ss.17A–17R and ss.66A–66L. The limit was 50 ML but was reduced to broaden uptake of the scheme.

[44] WIA ss.6–7; but note that the general provisions in WIA Part II regarding appointments, standard conditions etc. also apply.

[45] WIA s.17A.   [46] WIA s.66A.   [47] WIA s.17E.

[48] 'Water Supply Licences' see https://www.ofwat.gov.uk/competition/wsl/wsllicensees/.

[49] WIA s.66E.

[50] *Albion Water Ltd v Water Services Regulation Authority (formerly the Director General of Water Sevices) (Dwr Cymru/Shotton Paper)* [2006] CAT 23, [2008] EWCA Civ 536, [2008] UKCLR 457; *Albion Water Ltd v Water Services Regulatory Authority and Dwr Cymru Cyfyngedig, United Utilities Water PLC intervening* [2008] CAT 31. The case was an appeal by the company against a Decision by the DG under the Competition Act that *inter alia* the price being charged by Dwr Cymru to Albion Water was not an abuse of a dominant position. Extensive hearings before the Competition Appeals Tribunal, and some in the Court of Appeal, continued till 2013 when Albion Water were awarded their costs; *Albion Water Ltd v Dwr Cymru Cyfyngedig* [2013] CAT 16.

[51] Water Bill 2013 Part 1 Chapter 1; and see Chapter 3 on abstraction reforms.

[52] WIA ss.36A–36G.   [53] WIA s.18.   [54] WIA s.19.

[55] WA2003 s.48, inserting new ss.22A–22F into WIA.

[56] WIA ss.23–26.

[57] *Smartsource v IC and a Group of 19 additional water companies* [2010] UKUT 415 (AAC); *Fish Legal v IC* [2012] UKUT 177 (AAC).

[58] *Fish Legal (and another) v The Information Commissioner, United Utilities, Yorkshire Water and Southern Water* C-279/12. In the other jurisdictions, freedom of information laws, and in Scotland environmental information rules, would apply to public suppliers; Promotion of Access to Information (SA) Act 2000; Right to Information (Qld) Act 2009; Freedom of Information (Scotland) Act 2002; Environmental Information (Scotland) Regulations SSI 2004/520. In England until now, only the regulator, as a public body, has been subject to the Freedom of Information Act 2000. At least in Scotland, these rules can also be used to force disclosure of PPP contracts.

over 90% for sewerage).[59] The remainder are served by private water supplies and, usually, septic tanks. SW was set up in 2002 under the Water Industry (Scotland) Act 2002 (WISA).[60] This was not a consolidating act and many duties and functions remain in the Water (Scotland) Act 1980[61] and the Sewerage (Scotland) Act 1968.[62] The ownership model is therefore very different from England, but the broad regulatory structure is the same; Ministers set the policy context, SEPA authorises abstractions and discharges, and there is a separate Drinking Water Quality Regulator.

The Scottish Government is committed to retaining SW in the public sector. The Scottish water industry was not divested; this was proposed in 1992, but there was a public campaign against it and the Westminster Government instead restructured the industry, moving it away from the then regional councils to three regional water authorities.[63] A shift away from municipal delivery was seen as having several advantages, allowing a focus on the service not compromised by a variety of competing functions, and facilitating better accounting and asset management; this approach has been seen in many jurisdictions in recent years, including Queensland.

There have been several initiatives to increase market involvement in water services. From 1996 to 2002 there was a significant tranche of private funding for investment in wastewater treatment plant through the Private Finance Initiative, a form of PPP using BOO schemes. These were designed to provide capital investment in excess of that available via government borrowing, and in turn were required to implement the EU Directives on Urban Waste Water Treatment (UWWTD)[64] and Bathing Waters.[65] The UWWTD will be considered further below; overall, these schemes have cost more than they would through public borrowing and are unlikely to be used again for water in Scotland.

The other initiative has been limited market liberalisation, introduced for retail services only and for business customers only, under the Water Services (Scotland) Act 2005.[66] This Act replaced an individual Commissioner with a five-person Water Industry Commission for Scotland (WICS), and importantly gave the WICS the power to set charges, not just to advise Ministers; this is unusual in the public sector. The 2005 Act also established a licensing regime for new entrants. It created offences relating to putting water or wastewater into SW's systems, effectively prohibiting common carriage, and differentiated SW's 'core'

functions, which in turn enabled its commercial retail activities to be separated out into a competitive entity, Scottish Water Business Stream, which comes within the new licensing regime.[67] SW Wholesale is a supplier of last resort.[68]

As there are a relatively small number of business customers in Scotland, and the liberalisation only applies to retail services (billing, and other services such as water efficiency advice, but not distribution of water itself), it is a small-scale experiment. However, it has been successfully implemented and the WICS' view is that its existence is driving Business Stream to be more effective and provide a better service to businesses that have not switched.[69] As England proposes much more ambitious liberalisation, but including retail competition, the experience in Scotland should be of use and interest. Importantly, the access pricing for wholesale supply is public and determined as part of the price setting process, which creates transparency and avoids manipulation.

5.2.3 Queensland

In Queensland, recent reforms have been extensive and have included creating, and then disbanding, several institutions. Given the population size, the legal and institutional framework is very complex, though it does reflect the extensive land area, with a dispersed rural population and one highly urbanised conurbation. Queensland, like South Africa and many other countries, has a disaggregated system, and the distribution/retail service is managed differently in urban and rural areas, and differently again in Brisbane and the SEQ region. Outwith SEQ, services may be provided by local government or by rural water authorities. Some provision for bulk supply is still located in the QWA, which seems sensible as it provides for resource management. Although most regulatory functions for water services, both bulk and distribution/retail, rest with DEWS, DNRM also has some functions (relating to rural water authorities) and so do the Departments of Local Government and of State Development, Infrastructure and Planning.

Bulk water is principally provided by SunWater (for rural supply, especially irrigation), and by SEQWater (the Queensland Bulk Water Supply Authority) in that region. In 2006/7, there were extensive reforms to address the long-term drought. These established the Queensland Water Commission, subsequently abolished, and made new provision on supply and demand in the QWA.[70] As noted in Chapter 3, three bulk supply agencies were created, along with the office of grid manager, as part of an ambitious plan for security of supply. LinkWater was responsible

[59] Scottish Executive (2001).
[60] Water Industry (Scotland) Act 2002 asp.3 (WISA).
[61] Water (Scotland) Act 1980 c.45 as amended.
[62] Sewerage (Scotland) Act 1968 c.47 as amended.
[63] Local Government (Scotland) Act 1994 c.39; and see Hendry (2003).
[64] Directive 1991/271/EEC.
[65] Directive 1976/160/EEC, now 2006/7/EC.
[66] Water Services etc. (Scotland) Act 2005 asp.3.

[67] Business Stream must be fully ring-fenced and SW may not subsidise its activities; Water Services (Intra-Group Regulation) Direction 2006.
[68] Water Services etc. (Scotland) Act 2005 s.17.   [69] WICS (2011).
[70] Water Amendment (Qld) Act 2006 No.23, inserting new Chapter 2A into the QWA.

for the grid; WaterSecure, for recycled and desalinated water; and SEQWater as the bulk supply authority.[71] Under the current state government, and after the drought had broken, the Commission was disbanded, and the three agencies merged into SEQWater.[72] The current provision in the QWA enables setting 'level of service' objectives, nominating specific providers who must then establish water security plans,[73] and a Ministerial Code for bulk supply.[74] As ever, water is about politics – and money. The schemes in response to drought were expensive, and once the drought ended, a deregulatory government could stop these measures, reduce costs and lower prices.[75]

The remaining provision in the QWA relevant to services is for the water authorities, either category 1 or category 2. The former are larger and have more complex institutional requirements,[76] and there are only two, Mount Isa and Gladstone; it is not a coincidence that these are associated with the sites of major extractive industries, and they also provide bulk supply to the local governments concerned. The category 2 water authorities are generally irrigation boards, drainage boards, or provide water for stock and domestic use.[77] If they provide domestic water services they must also be registered as service providers under the Water Supply (Safety and Reliability) Act 2008 (Water Supply Act),[78] which took dam safety and service provider obligations out of the QWA, and also provides for drinking water quality standards and the use of recycled water.

Most water and sewerage services in urban areas are provided by local governments, regulated generally under the Local Government Acts.[79] These provide *inter alia* for a principle of 'competitive neutrality', applying where a local government is providing a service that could be provided by the private sector, and it is a 'significant business activity'.[80] The Local Government Regulation then provides more detail; if the service is above a certain threshold, it may require full cost pricing, 'commercialisation', or full 'corporatisation' via a separate entity.[81] If there are any Community Service Obligations (where a service is subsidised) then these must be accounted for as

revenue under these structures. There is a specific threshold for water and sewerage services,[82] and for the use of two-part tariffs for these services.[83] Complaints over competitive neutrality are made to the Queensland Competition Authority. These provisions were driven by the Federal COAG policy on competition, and now the NWI.[84] The COAG reforms require cost recovery even for rural water services – at least, management costs, and where practicable a return on capital. Urban water services should have consumption-based pricing and full cost recovery, with transparency over any retained cross-subsidies.

There are currently 77 local governments in Queensland, and despite some recent rationalisation, many are very small. These, along with some water authorities, will be service providers regulated under the Water Supply Act. In SEQ, following from the bulk supply reforms, in 2009 three combined distributor-retailers were established. The principal legislation is the SEQ Water (Distribution and Retail Restructuring) Act 2009 (SEQ Water Act), which provides for their governance and makes some parallel provision to the Water Supply Act.[85] One of the three, Allconnex, was subsequently disbanded and its functions returned to the three constituent councils;[86] but their service provision is still managed under the SEQ Water Act. This does create an additional layer of complexity.

5.2.4 South Africa

In South Africa, the problems and issues in water services are of a different order of magnitude. Prior to 1994, 75% of the population subsisted on just 13% of the land, and of a population of 41 million, 12 million had no water supply and 21 million were without sanitation.[87] Legislative reform began with the 1994 White Paper,[88] which adopted the basic principle of 'some for all not all for some'. A later variation became 'some for all forever', adding the sustainability dimension. Other principles in the White Paper included water as a human right; water services to be demand driven and community-development based; equitable regional allocation; water as having an economic value; the need for integrated development and environmental integrity; and the user pays and polluter pays principles. In 2003 the White Paper was reviewed and a new Strategic Framework for Water Services was approved.[89] This gave new figures on access to

---

[71] SEQ Water Restructuring (Qld) Act 2007 No.58.

[72] SEQ Water (Restructuring) and Other Legislation Amendment (Qld) Act 2012 No.39.

[73] QWA ss.350–358; these are alternatives to drought plans under the Water Supply (Qld) Act (below).

[74] To replace the prior 'Market Rules' for the Grid; QWA ss.360M–360T.

[75] In Queensland, bulk prices are set by the state, and retail prices by the distributors (below).

[76] QWA Chapter 4.

[77] There are currently 51; see 'Water Authorities' http://www.nrm.qld.gov. au/water/regulation/water_authorities.

[78] Water Supply (Safety and Reliability) (Qld) Act 2008 No.34 (Water Supply Act).

[79] Local Government (Qld) Act 2009 No.17; City of Brisbane (Qld) Act 2010 No.23.

[80] Local Government Act ss.43–48; City of Brisbane Act ss.47–52.

[81] Local Government Regulation 2012 SL No.236 Chapter 3 Part 2.

[82] Local Government Regulation Reg.19; AUD13.3 million, compared to AUD8.9 million for other services; special provision is made elsewhere for Brisbane and the SEQ region, below.

[83] Local Government Regulation Regs.40–41; and see further below on tariffs.

[84] COAG (1994, 1995, 2004); and see also Chapter 3.

[85] South East Queensland Water (Distribution and Retail Restructuring) (Qld) Act 2009 No.46 (SEQ Water Act).

[86] South East Queensland Water (Distribution and Retail Restructuring) (Qld) Act 2012 No.1; and see Baumfield (2012).

[87] Abrams (1996). [88] DWAF (1994). [89] DWAF (2003).

services from the 2001 census – of 44.8 million people, more than 10% had no access to safe water supply and a further 14% did not have a basic service as defined (see further below); some 40% did not have adequate sanitation facilities. In 2010, from a population of 50 million, WHO reported that some 91% had access to improved water supply and 79% had access to improved sanitation.[90] In its most recent annual report, DWA gives a headline figure for water supplied at 95%.[91] However, there are data at household level in the 2011 census for both services, and it reports somewhat differently, indicating that some 8% have no access to piped water, 85% meet the standard for accessibility, and perhaps another 7% have a service that does not meet the standard.[92] The census does not report on water quality as such, but 80% of supply is from a water supply scheme; most of the remainder is from sources that might or might not be 'improved' according to WHO definitions.[93] For sanitation, almost 70% have a facility definitely within the 'improved' definition, and some 5% have no facility at all.[94] The data, like the institutional arrangements, are complex and difficult to reconcile; but rural areas, informal settlements, and households headed by black Africans, are least well served. Although much progress was made in the early years, that progress has now slowed and in early 2014 there were protests, sometimes violent and including some deaths, at failure to provide services, including water, to the poorest.[95]

Constitutional responsibility for water services rests with the municipalities, supported by national and provincial governments.[96] The DWA is responsible for regulation of water services, including drinking water quality, bulk supply including infrastructure and pricing, and abstractions and discharges under the NWA. However, municipalities report to the Department of Cooperative Governance and Traditional Affairs (CoGTA), which funds infrastructure development;[97] and the Department of Human Settlements (DHS) has taken responsibility for providing basic sanitation.[98] Sanitation is closely tied into housing, and is especially relevant to slum housing, but there will always be impacts on water, especially groundwater, from the disposal of solid wastes and sludges, as well as effluents from waterborne systems.

There is a recognised need for more clarity around responsibilities for sanitation; the move to DHS of basic sanitation provision was problematic,[99] but equally there is a perception that DWA is much more focused on water supply than on sanitation.[100] Most recently, a Ministerial Task Force was established under DHS to examine future provision. An extensive report analyses the law and policy, the institutional framework, the failure to meet policy goals including eradication of bucket systems, and various 'malpractices' in delivery.[101] Whilst this recognises the DWA's overall supervisory role, it may also reflect tensions between departments. It recommends better coordination, but also establishing a new Agency that would take budgets and functions from all the relevant departments, and enable a strengthened focus on sanitation as such, which is certainly needed. The regulatory and institutional environment remains complex and subject to change.

The DWA is currently proposing integrating the water services legislation into the National Water Act, arguing that this will allow greater consistency, and make it easier to manage water through the value chain.[102] Although it is unusual, and not necessarily desirable, to provide for water resources and water services in the same legislation, the proposal also reflects the need for institutional clarity.

In South Africa also, there is vertical disaggregation between bulk suppliers and WSPs, and different structural provision in urban and rural areas. The Water Services Act 1997 (WSA)[103] creates the institutional structure and legal framework, and identifies various water services institutions, including water services authorities, which are municipal authorities with responsibility for access to water services; Water Boards, which supply bulk untreated and treated water on a commercial basis to other institutions; and water services providers, who provide services to consumers or other water services institutions. These are also likely to be local government, but may be Water Boards, or potentially private sector bodies. These providers will have a contract with the relevant Water Services Authority. Within the nine provinces there are 237 municipalities with water services functions; in some cases DWA is the Water Services Authority, although it has devolved this to the municipalities (at different levels) wherever there is capacity. In some localities, given the problems with municipal supply, Water Boards are effectively providing services at regional level; the intention under the current policy review is to rationalise the Water Boards and create rather fewer Regional Water Utilities, better able to manage large infrastructure.[104]

[90] WHO/UNICEF (2012).   [91] DWA (2013c) p.4.
[92] Statistics South Africa (2011) Table 3.13.
[93] Statistics South Africa (2011) Figure 3.5.
[94] Statistics South Africa (2011) Table 3.17.
[95] See, e.g., Miranda (2014) 'South Africans Protest at Lack of Basic Services', The Real Agenda News (online); Sky News (13 February 2014) 'South Africa Water Protesters "Shot by Police"' (online); DWA (2013d).
[96] Constitution of South Africa s.156 and sch.4B.
[97] See CoGTA (2013) sub-programme 3.4. For two years, 2010–12, CoGTA reported on achievements towards the targets for water services provision, replacing much fuller reporting from DWA.
[98] See DHS (2013) pp.59–60.

[99] See, e.g., AMCOW (2011), Tissington (2011).   [100] DHS (2012).
[101] DHS (2012).   [102] NWPR 2013.
[103] Water Services Act 1997 No.108 (WSA).
[104] NWPR 2013 section 3.2; NWRS2 section 8.1.3.

The legislation does allow for the involvement of the private sector in water services provision, through contracts or joint ventures with the water services authorities,[105] but not divestiture; any contracts to provide water services must be of limited duration.[106] There is an explicit policy preference for public sector suppliers: 'a water services authority may only enter into a contract with a private water services provider after it has considered all known public water services providers which are willing and able to perform the relevant functions'.[107] This type of provision may be useful to states seeking to minimise public concern over PSP.

There are a small number of long-term concessions in South Africa, including in the Ilembe District (formerly Dolphin Coast) and Nelspruit. In Johannesburg the municipality formed Johannesburg Water and entered a management contract with a joint venture including Suez Environnment, but this was not renewed. Thus the legislation makes limited provision for private sector involvement in various forms, but generally services are provided by public authorities of different types. There are extensive provisions on delivering services, with various structures, within and across municipalities, under the Municipal Systems Act.[108]

5.3 CONSTITUTIONAL RIGHTS AND HUMAN RIGHTS

The international debate around the human right to water has been noted. Absent a treaty, and whether or not a customary right exists at international law, states may choose to enshrine a constitutional right to water as a backdrop to any specific industry duties. The UK jurisdictions do not have a formal written constitution, though, as discussed in Chapter 3, some domestic legislation may have constitutional effect, including the Human Rights Act, implementing the European Convention on Human Rights. The UK Government has been reluctant to accept a human right to water, and especially to sanitation, and abstained at the vote in the UN General Assembly.[109] Although both federal Australia and Queensland have written constitutions,[110] these do not contain a Bill of Rights; Australia also abstained from the General Assembly resolution in 2010. Only South Africa makes constitutional provision, mandating 'access to … sufficient food and water'; the state 'must take reasonable

legislative and other measures, within its available resources, to achieve the progressive realisation of each of these rights.'[111]

One benefit of a constitutional right is that it opens up litigation in the Constitutional Court, and there has been some case law, most famously the *Mazibuko* litigation.[112] Here, reversing the decisions of the High Court and the Court of Appeal, the Constitutional Court held that there was a duty on Johannesburg City Council to provide the 25 LPD required under the free basic water policy (see further below), but did not enforce the policy aspiration to extend this to 50 LPD.[113] The right was progressive, but councils had other obligations, and if the Court required them to provide 50 LPD, or indeed 42 LPD as the Court of Appeal had ordered, then other services would suffer. The decision was disappointing to those arguing for a more expansive provision, but it is an excellent example of the need for specific sectoral duties. The Court did not consider it to be their role to direct in detail how councils provided services as long as the constitutional provision was met, and focused rather on the progressive realisation; but courts will enforce any minimum legal requirement within that constitutional right.

The right to sanitation is not express in the Constitution, though it can be implied into the right to adequate housing, as well as rights to dignity and privacy; cases on sanitation in the Constitutional Court have been argued on this basis.[114] The rights to basic services will be considered further below. In terms of service provision, both constitutional rights and high level duties can only be effective where there are clearly specified duties of supply, detailing the levels of service for water supply, sanitation and wastewater management.

5.4 DUTIES OF SUPPLY

In every jurisdiction there are high level provisions, especially around conservation and water efficiency, specific to service providers (or their regulators); these will be considered below in relation to conservation and demand management. This section will consider duties of supply, including responsibility to provide connections or expand networks; the supply of basic

---

[105] WSA s.19.   [106] WSA s.22.   [107] WSA s.19(2).

[108] Municipal Systems Act 2000 No.32 (MSA) as amended, MSA 2003 No.44. This Act will be considered in more detail in relation to economic regulation and business planning.

[109] UN General Assembly (2010a).

[110] Constitution of Australia Act 1900; Constitution of Queensland Act 2001 No.80.

[111] Constitution of South Africa s.27. The Constitution also provides for a clean environment, s.24; housing, s.26; and health, food, and social security, s.27.

[112] *Mazibuko v City of Johannesburg* (39/09) [2009] ZACC 28. There has been extensive analysis of the provision of water and sanitation in South Africa, including analysis of case law; see, e.g., Algotsson and Murombo (2009); COHRE (2008); Tissington (2011); Mjoli (2009).

[113] The aspiration is found in the Strategic Framework, DWAF (2003).

[114] *Nokotanya and Others v Ekurhuleni Metropolitan Municipality and Others* (31/09) [2009] ZACC 33; *Beja and Others v Premier of the Western Cape and Others* (2011) Case No:21332/10; and see further below.

services; standards for drinking water quality and wastewater removal and treatment; and other customer service standards.

In England, there are general duties on undertakers to supply water and sewerage systems that appear unqualified,[115] but in fact are limited by the more general duties of the regulator and the Secretary of State. In Scotland, there are duties to supply water and sewerage services, but only where this can be done at 'reasonable cost'.[116] In Queensland, there is an obligation to provide services 'to the greatest practicable extent', to customers in a defined service area.[117] In South Africa, there is a universal service obligation: 'everyone has a right of access to basic water supply and basic sanitation';[118] and progressive requirements to supply more extensive service.[119]

5.4.1 England

In England, due to the WSPs being fully divested, there is a series of high level duties that apply to the regulators – the Secretary of State and OFWAT (the Authority). The first of these, under the WIA as enacted, was to secure the carrying out of the functions of water undertakers, and the second was to ensure that the companies were able to make a reasonable return on their capital. Secondary to these were the duties to protect customers, to promote economy and efficiency, and to facilitate competition. However, currently the furthering of the 'consumer objective' is the first duty.[120] This is further defined as protecting their interests 'wherever appropriate by promoting effective competition'.[121] There is a further requirement to give special regard (but not exclusively) to certain groups of consumers including the sick and disabled, pensioners, those with low incomes and those in rural areas, reflecting concerns over the social agenda. They must also 'promote economy and efficiency', 'contribute to the achievement of sustainable development', and 'have regard to principles of best regulatory practice'.[122] Further reforms proposed under the current Water Bill, including a duty on OFWAT to 'secure resilience',[123] reflect concern over climate change and other pressures on the resource.

The WIA requires the provision of water supply, made available to 'persons who demand' such supply,[124] and undertakers must make connections for domestic supply upon service of a notice, but with the costs borne by the customer.[125] There is separate provision for supply for non-domestic use, by agreement.[126] For sewerage,

there is a duty to provide 'effectual drainage',[127] and also to provide a public sewerage system where one has been requisitioned, again on payment of the cost by the customer spread over 12 years.[128] Although these look like unqualified duties, they are balanced and indeed superseded by the general duties above, particularly to ensure that the interests of all customers are protected and that the water companies can make a return on their capital. A House of Lords decision confirmed that the duty to provide effectual drainage is not an absolute duty, but is subject to the wider policy context, which might prioritise other areas for investment where more customers would benefit.[129] Further, this case, which included a human rights argument regarding the right to enjoyment of property, clarified that the appropriate means of enforcing these duties is through OFWAT and not by private law actions.[130]

5.4.2 Scotland

In Scotland, SW has a duty to supply both water and sewerage services where this can be done at 'reasonable cost'.[131] This is understood to exclude remote rural areas, where there is no public network provision, but there are also issues around connections for new developments where networks may be at capacity, but capable of extension. In practice and in the past, SW defined reasonable cost as somewhere in the region of £1200 per connection; if a new connection cost less than this, it was subsidised.[132] This was a matter of concern to the then Government, so now legislation provides for a definition of reasonable cost by regulation.[133] The duty to 'contribute to sustainable development' in Scotland is placed on SW, and not on the regulator.[134]

The duty to supply water is for 'domestic purposes' as defined,[135] and there is provision for non-domestic supply by

---

[115] WIA s.37, s.45, s.94.

[116] Water (Scotland) Act 1980 s.6; Sewerage (Scotland) Act 1968 s.1.

[117] Water Supply Act s.164.   [118] NWA s.3.   [119] NWA s.11.

[120] WIA ss.2A–2E.

[121] WIA ss.2–2B, reflecting provision made for other services in the Utilities Act 2000 c.27, s.9, s.13.

[122] WIA s.2(3).   [123] Water Bill cl.22.   [124] WIA s.37.

[125] WIA s.45.

[126] WIA s.55; disputes are determined by the Authority, s.56.

[127] WIA s.94.

[128] WIA s.99. Although almost all of England is served by mains drainage, there has been recent action to transfer the ownership of some lateral drains and private sewers from property owners to the WSPs, where these were draining to a public mains sewer and treatment works. Although expensive, this was considered to be an effective way to ensure the best use of the networks and to avoid deterioration where private owners failed to maintain; WIA s.105A–105C, Water Industry (Schemes for Adoption of Private Sewers) Regulations SI 2011/1566.

[129] *Marcic v Thames Water PLC* [2004] 1 All ER 135.

[130] Although the action did succeed where Mr Marcic was concerned, as improvements were made following the judgment in his favour in the Court of Appeal.

[131] Water (Scotland) Act s.6, Sewerage (Scotland) Act s.1.

[132] Scottish Executive (2001).

[133] WEWS s.29, and the Provision of Water and Sewerage Services (Reasonable Cost) (Scotland) Regulations SSI 2006/120.

[134] WISA s.51; in part this is because SW is a public body, but the WICS has also resisted such a duty when it has been proposed, as they consider that is not the appropriate balance for the economic regulator.

[135] Water (Scotland) Act 1980 ss.6–7. Domestic purposes include drinking, washing, cooking, central heating, baths below a certain capacity and some business use where the premises are mainly residential.

commercial agreement. The practice in Scotland is for local planning departments to consult SW and refuse planning permission for developments where networks are already at capacity, but the other side of this is that SW has been seen as holding up development and this was an issue in the run-up to the 2006–10 price review.[136] The rules on reasonable cost were introduced in the context of policy initiatives to separate out the strategic elements of the public system, for which SW should be responsible, and the local and householder elements for which developers (or householders) should be responsible.[137]

5.4.3 Queensland

In Queensland, as noted, there are two parallel regimes. The Water Supply Act applies to 'service providers', which include local governments and water authorities, and these must be registered;[138] but the distributor-retailers are also service providers and some requirements of the Water Supply Act also apply to them.[139] Duties to supply emerge once an area has been declared a service area by a local government,[140] with a designated supplier. Where the WSP is not local government, the entity must first agree to the designation. Once designated, the WSP must ensure 'to the greatest practicable extent' that all premises or groups of premises can be separately connected.[141] 'Reasonable costs' of giving access may be recovered.[142] There is no duty to supply if 'physical constraints' prevent supply at satisfactory pressure, unless the householder supplies adequate storage and pumping facilities to overcome this.[143] If the owner of premises asks for a connection, he may have to do works and pay the connection fee, and WSPs can require work to be done to enable a connection.[144] For the distributor-retailers, service areas are established by their participating local governments, until an appropriate plan is in place.[145] The SEQ Water Act also provides for a Code, which *inter alia* gives advice to householders and small businesses as to connection requirements.[146]

5.4.4 South Africa

In South Africa, the critical focus is on the provision of basic services. The first object of the Act is to provide basic water supply and basic sanitation,[147] and everyone has a right to these services.[148] Authorities must take 'reasonable measures' to

realise these rights,[149] and this has priority over provision of other water services.[150] Although the 1994 White Paper expected all consumers to pay for their water services, in 2001, the Government introduced the Free Basic Water (FBW) programme, to provide a minimum level of service to the poorest citizens.[151] Other basic services, including sanitation (FBSan), may also be provided free to those who cannot pay; this is a matter for the municipalities when setting tariffs (below).

The WSA also places a general duty on water services authorities, to 'progressively ensure efficient, affordable, economical and sustainable access to water services' to 'all consumers or potential consumers' in their area, but qualified by availability of resources, equity, the duty to pay charges, the duty to conserve resources, the physical environment, and the right to limit or discontinue supply.[152]

Further detail on connections is not provided in the WSA, and needs to be made by the municipalities. However, DWA has issued model bylaws which municipalities can adapt,[153] and these do provide procedures for connections, including allocation of responsibility for service pipes etc. Water services authorities have a duty to draw up bylaws,[154] as well as Water Services Development Plans, under the WSA.[155] The model bylaws also provide a framework for the delivery of FBW (and FBSan), depending on whether the authority is supplying these services to all its citizens or only to the indigent poor.[156] Because of the special features of the basic services provision in South Africa, these will be addressed together, and separately from the other duties of supply and service standards.

5.4.5 Basic services in South Africa

In South Africa, basic water supply and basic sanitation are defined in the Act by reference to prescribed minimum standards necessary for the 'reliable supply of a sufficient quantity and quality of water' and 'the safe, hygienic and adequate collection, removal, disposal or purification' of human waste.[157] Water supply should be sufficient for personal hygiene, and there is reference to informal households. Prescribed standards were then defined in regulations (the Standards Regulations),[158] and these definitions were revised under the Strategic Framework, to specify both the facility and the service.[159]

---

[136] Scottish Executive (2004a); Scottish Executive (2004b).

[137] Scottish Executive (2005b); Scottish Executive (2006b).

[138] Water Supply Act s.20.   [139] SEQ Water Act Chapter 2A.

[140] Water Supply Act s.161.   [141] Water Supply Act s.164.

[142] Water Supply Act s.165.   [143] Water Supply Act s.166.

[144] Water Supply Act ss.167–168.

[145] SEQ Water Act ss.53AP–53AQ. The business planning system, which is different for the distributor-retailers, will be considered below.

[146] SEQ Water Act Chapter 4 and DEWS (2013).   [147] WSA s.2.

[148] WSA s.3

[149] WSA s.3.   [150] WSA s.5.

[151] DWAF (2003); DWAF (2007c); and see Muller (2008).

[152] WSA s.11.   [153] DWAF (2005).   [154] WSA s.21.

[155] WSA s.12. The business planning process will be considered below.

[156] And see COHRE (2008) for a critique of the delivery of services, and the bylaws, in 15 municipalities.

[157] WSA s.1.

[158] WSA ss.9–10; Regulations Relating to Compulsory National Standards and Measures to Conserve Water No.509 of 2001 (Standards Regulations).

[159] DWAF (2003) Table 2. Ideally these new definitions should also be provided in regulations.

For water supply, the 'facility' is 25 LPD, or 6000 L/month per connection (an assumption of eight persons per household), with minimum flow of 10 L/minute, and within 200 m. The 'service' is availability for 350 days per year, along with information on water use and hygiene. For sanitation, the 'facility' is defined in terms of its safety, reliability, privacy, and ability to minimise disease and enable removal and treatment of waste. The service is accessibility to the household and 'sustainable operation' of the 'facility', including removal of waste and communication of good hygiene and related practices. The Strategic Framework recognised a need for a progressive approach. It also considered the needs of the peri-urban poor, where there are issues of land tenure as well as ability to pay. In many countries, informal settlements are barred from accessing networked services even if such exist close by, as legal tenure is often a prerequisite for a service contract.

For the basic water service, the 25 LPD figure is the subject of ongoing debate; the WHO gives a figure of 20 LPD as the absolute minimum for drinking, cooking and hygiene, although a general figure of between 20 and 40 LPD is often given.[160] In a much-cited article, Gleick recommends 50 LPD.[161] The Strategic Framework recommends an aspiration of 50 LPD; but, as discussed, in the *Mazibuko* litigation an entitlement above 25 LPD was not imposed by the Constitutional Court. Any higher figure would also need to be incorporated into the Reserve for basic human needs, currently calculated on 25 LPD. There is also a debate as to the assumption of eight persons per household, as that is more likely to be an underestimate for the poorest.

Sanitation is always the poor relation of water supply: more expensive, less politically attractive, often taboo. There has been a series of policy initiatives recognising the particular problems with sanitation, going back to a separate White Paper in 2001.[162] Non-waterborne sanitation has been encouraged, in recognition of the scarcity of water, but in urban areas especially there are significant difficulties with the management of the resultant waste.[163] Reports for the Water Research Commission and the Socio-Economic Rights Institute both consider there is a discrepancy between this White Paper and the Strategic Framework that has led to municipalities taking the view that emptying facilities is a household obligation.[164] That seems an unrealistic expectation regardless of the level of education provided, and municipalities (or some public authority) will surely need to take responsibility for the removal of the waste if onsite sanitation of any type is to be a long-term solution. The Department has agreed that it should be clarified.[165]

The South African courts have examined rights to sanitation in the context of housing policy; in *Beja*, the Court held that open toilets did breach the right to privacy, but did not rule on the right to sanitation as such.[166] In *Nokotanya*, the Court upheld the Council's arguments that they had complied with the Housing Code, and did not admit new arguments under the Water Services Act.[167] The decisions are each understandable, but neither goes very far in improving the lot of those without even a basic service.

5.4.6 Drinking water quality

There are WHO guidelines on drinking water quality, which many states adapt as a basis for national standards.[168] An important feature of the latest version is the emphasis on prevention of contamination of supply sources, using water safety plans and linked to catchment protection as discussed in Chapters 2 and 4.

In the UK, the first EU Drinking Water Quality Directive[169] set mandatory technical standards for drinking water for the first time. The current (second) Directive is based on, and complies with, WHO guidelines. In England, transposition of the original Directive caused particular problems and resulted both in domestic judicial review,[170] and in enforcement actions by the European Commission in the European Court of Justice, for failure to transpose.[171] In both UK jurisdictions there is separate regulation of the standard of drinking water, and in England the Drinking Water Inspectorate has wide powers of investigation under WIA.[172] Generally though, in England drinking water quality is very high, with compliance for most WSPs above 99%. This reflects the huge investment made in England to improve treatment and comply with the Directive. The need for that investment was one driver for divestiture.

In Scotland, as in England, prior to the first Directive, the only statutory requirement was that drinking water be 'wholesome'.[173] New provisions were added to the 1980 Act enabling

---

[166] *Beja and Others v Premier of the Western Cape and Others* (2011).
[167] *Nokotanya and Others v Ekurhuleni Metropolitan Municipality and Others* (2009).
[168] WHO (2004).   [169] Directive 1980/778/EEC.
[170] *R v Secretary of State for the Environment ex parte Friends of the Earth* [1995] EnvLR 11.
[171] *European Commission v UK* [1992] ECR I-6103 C-337/89; *European Commission v UK* [1999] ECR I-2023 C-340/96. The judicial review failed, in that although water supplied was not 'wholesome' by the new definition, no loss or harm had resulted. However, the European Court of Justice found against the UK, as the Government's preferred mechanism for enforcement was not s.18 orders, but rather accepting the undertakings of the companies under s.19. The system of undertakings was not an adequate legal framework for compliance with EU law.
[172] WIA s.86.
[173] Water (Scotland) Act 1980 s.6; and see *McColl v Strathclyde Regional Council* 1983 SC 225, where 'wholesome' was defined as 'pleasant and fit to drink', which is little use to an engineer designing a treatment process.

[160] WHO/UNICEF (2006).   [161] Gleick (1996).
[162] DWAF (2001), DWAF (2008b). For a comprehensive review, see Tissington (2011).
[163] COHRE (2008).   [164] See Tissington (2011), Mjoli (2009).
[165] DWA (2010a).

the Secretary of State to make regulations for the standard of drinking water,[174] and creating new enforcement mechanisms. Existing powers to issue default orders[175] were supplemented by powers to make enforcement orders.[176] As well as the statutory duty to supply wholesome water, there is also a criminal offence of supplying water unfit for human consumption.[177] Drinking water quality is now regulated by the Drinking Water Quality Regulator; the annual reports indicate sustained improvements in recent years.[178]

The second Drinking Water Quality Directive places more emphasis on monitoring at the tap rather than the end of the WSP's pipes. It applies to private water supplies as well as the WSPs, with separate implementing legislation, and here local authorities have the duty to monitor and enforce water quality.[179] Local authorities also have general functions in both jurisdictions to liaise with WSPs and keep themselves informed about (public) water quality in their areas, as any serious breach may have public health consequences.

In Queensland, as in other Australian states, there is no single mandatory standard for drinking water quality, but there are Federal guidelines.[180] Again, these are based on WHO guidelines for safe water, but this should be 'aesthetically pleasing', whereas WHO guidelines suggest it be 'acceptable'.[181] There is recognition that mandatory technical standards may be inappropriate for every part of every state, and especially, disproportionately expensive for rural areas with small populations.[182] The Guidelines therefore suggest procedures for community consultation and participation, to make appropriate trade-offs. Small communities in every jurisdiction are more likely to use private sources such as boreholes, and be responsible at community level for maintaining the source and treating the water. The founding principle for small communities everywhere is testing more frequently for a narrow range of key parameters rather than less frequently for a larger range.

In Queensland, service providers generally should include their criteria for drinking water quality in Drinking Water Quality Management Plans (DWQMPs).[183] The SEQ Water and Sewerage Code refers to either DWQMPs or the Australian guidelines as a basis for relevant service standards.[184] As the Water Supply Act has taken some time to implement, especially for smaller WSPs, a separate notification system has also been

implemented by DEWS for authorities without a DWQMP, requiring them to report on their monitoring and on any incidents.[185] These specifically include exceeding the parameters in the Australian guidelines.

In most countries, departments of public health have a role in drinking water quality, and Queensland Health is responsible for the relevant regulations as well as for the fluoridation of water supplies. It is an offence to supply water that is 'unsafe' under the Public Health Act, and improvement notices can be served.[186] The Public Health Regulation sets a minimum standard for e-coli as a measure of faecal contamination, for fluoride, and for recycled water (below).[187] In Queensland, as in England and Scotland, there are specific powers to add fluoride to water.[188] It has been done in Queensland and in some English regions, but not in Scotland to date. As this is politically controversial, and may be ultra vires, a specific power is inevitably required.[189]

During the drought, one initiative in Queensland was the reintroduction of purified treated wastewater into the drinking water supply system. Although the relevant institution, Water-Secure, has now been wound up, its functions have moved to SEQWater and the Water Supply Act makes detailed provision for the authorisation of the process,[190] whilst the Public Health Regulation provides the required standards, including for use as irrigation water.[191] In a time of increasing pressure on the resource, the reuse of treated wastewater, and its return to the freshwater system, rather than discharge into coastal waters, is likely to be a focus in many countries; it will be considered below in relation to conservation.

In South Africa, the same drinking water quality standards should apply to both the basic service and full urban supply. The Standards Regulations require a sampling programme,[192] narrated in the water services authority's development plan, and making reference either to national drinking water standards[193]

---

[185] 'Drinking Water Quality Management Plans' see generally http://www.dews.qld.gov.au/water-supply-regulations/drinking-water/drinking-water-quality-management-plans.

[186] Public Health (Qld) Act 2005 No.48 Chapter 2 Part 5A; 'unsafe' water is defined as likely to cause harm. There is also a specific prohibition on using lead in water fittings and a duty on householders to remove any lead of which they are aware; Part 6. Lead is also a problem in some parts of the UK, but such a duty has not been imposed there.

[187] Public Health Regulation 2005 SL No.281, Part 6A and Sch.3A.

[188] Water Fluoridation (Qld) Act 2008 No.12; WIA Part III Chapter IV; Water (Fluoridation) Act 1985 c.63.

[189] The issue was tested in Scotland, in McColl v Strathclyde Regional Council (1983). It was held that whilst SRC were not in breach of their duty to provide wholesome water, they would be exceeding their powers, as the water was wholesome without the fluoride. The case led to the Water (Fluoridation) Act 1985, but no fluoridation has ever taken place in Scotland.

[190] Water Supply Act Chapter 3.

[191] Public Health Regulation 2005 Regs.18AD–18AH and Sch.3A.

[192] Standards Regulations Reg.5.

[193] SANS 241: Specifications for Drinking Water; and see DWAF (2005a).

[174] Water (Scotland) Act 1980 s.76J, inserted by the Water Act 1989 sch.22; Water Supply (Water Quality) (Scotland) Regulations 1990 SI 1990/119 as amended.

[175] Water (Scotland) Act 1980 s.11.

[176] Water (Scotland) Act 1980 s.76E, making parallel provision to WIA s.18.

[177] Water (Scotland) Act 1980 s.76C.

[178] Water Industry (Scotland) Act Part 2; and see DWQR (2012).

[179] Private Water Supplies (Scotland) Regulations SSI 2006/209.

[180] NHMRC/NRMMC (2011).   [181] NHMRC/NRMMC (2011) p.7.

[182] NHMRC/NRMMC (2011) Chapter 4.

[183] Water Supply Act ss.94–95.   [184] DEWS (2013) Section 5.2.

or to national guidelines.[194] The Strategic Framework provides the same standard for potable water for basic supply.[195]

Further guidance was produced by DWAF in 2005.[196] Water treatment plant should be registered with the Department, which also manages an accreditation scheme for operators. Although Water Service Authorities will often be monitoring themselves (as they are usually also WSPs), they report to DWA as regulator. There are also powers for environmental inspectors to test drinking water.[197] Currently the Department is operating a 'Blue Drop' scheme, for drinking water assessment and compliance.[198] This takes a proactive and incentive-based approach, and encourages water safety planning as part of the overall Blue Drop scoring. It has reported significant improvements in water quality, and uses a web-based reporting system that allows consumers to see at a glance (assuming they have access and capacity) how their municipality is performing. In 2010, DWA issued a new Regulatory Strategy which makes compliance with drinking water standards one of three priorities.[199]

5.4.7 Sewage and wastewater treatment

Historically, waterborne sewage was discharged untreated, to sea or inland waters; this is still the case in many places. Treatment produces effluent, with its quality dependent on the level of treatment, and sludge, which becomes a solid waste problem; both contain high levels of nutrients, but also pathogens, heavy metals and other contaminants.[200] Wastewater and sludge may be used in various ways, with different levels of risk;[201] this is of increasing interest both in terms of the environmental consequences of inadequate disposal, and in the reuse of wastewater. Discharges of wastewater from sewers and treatment plant of different types, and discharges made directly from industrial processes, are regulated under pollution control legislation (Chapter 4).

In England, licensing of discharges from sewers or works is carried out by the EA under the EPR. There is an exemption for small domestic systems discharging less than 5 m^3/day,[202] and standard rules apply to discharges of 5–20 m^3/day.[203] In

Scotland, there is a similar tiered approach regulated by SEPA under the CAR. Small onsite systems (normally septic tanks) of less than 15 population equivalent are registered, to allow the regulator to identify any cumulative impacts; larger systems are licensed.[204]

In both Scotland and England, specific standards for wastewater treatment are set by the EU under the UWWTD[205] and implementing regulations.[206] The Directive required that 'collecting systems' be put in place for domestic and biodegradable industrial wastewater, and required 'secondary treatment' as defined in Annex I. As with the first Drinking Water Quality Directive, this necessitated major investment in treatment plant. The Directive continues to be problematic for both long-established and new Member States, although the most recent compliance report shows some improvement.[207] Although it is arguable that it has encouraged large capital projects and energy-intensive treatment systems, it has also stopped the practice (certainly in the UK) of discharging raw sewage to sea via long pipelines, or dumping the same from ships.

In Queensland, discharges from treatment plant etc. are regulated by the DEHP under the Environmental Protection Act and Regulation. Sewage treatment is an ERA that requires an authorisation,[208] but again the approval process is dependent on scale. Plant treating less than 21 population equivalent are covered by the Plumbing and Drainage Code,[209] and above that a sliding scale based on population equivalent determines the level of authorisation, by local government, the DEHP or the Department of State Development, Infrastructure and Planning.[210]

In South Africa, the NWA's system for integrated water use licences includes the licensing of discharges of waste and wastewater into watercourses.[211] Discharges are exempt only if they are made into a channel, conduit, etc. that is under the control of another person who will carry out appropriate treatment.[212] For the middle tier, the general authorisation makes provision for certain discharges, for onsite sewerage (septic tanks, soakaways, pit latrines etc.) and for the storage and reuse of wastewater for irrigation, within volumetric limits and as long as the effluent meets certain quality standards.[213] If the volume or density is above a certain limit the local authority for the area must register the use, rather than the owners or users of the system, which seems a very sensible provision. The general authorisation has

---

[194] DWAF (1996) Vol.1.   [195] DWAF (2003) para.6.3.2.

[196] DWAF (2005a, 2005b).

[197] National Health Act 2003 No.61 s.83, providing for environmental health investigations; 'municipal health services' are defined to include water quality monitoring (s.1).

[198] DWA (undated).   [199] DWA (2010a).

[200] See, in the UK, Sewage Sludge (Use in Agriculture) Regulations SI 1989/1263, implementing Directive 86/278/EEC (currently due for review, but proving problematic, under the Soil Strategy, European Commission (2006), European Commission (2006a)); in South Africa, DWAF (2006); in Australia, ANZECC/ARMCANZ (2004). In Queensland there is specific legislative provision for identifying waste as a resource, which would apply to biosolids; Waste Reduction and Recycling (Qld) Act 2011 No.31 Chapter 8.

[201] WHO (2006).   [202] EPR Sch.3 Part 2.   [203] EA (2010a).

[204] CAR Regs.7–8; SEPA (2013b).   [205] Directive 1991/271/EC.

[206] Urban Waste Water Treatment (England and Wales) Regulations SI 1994/2841, UWWT (Scotland) Regulations SI 1994/2842.

[207] European Commission (2013a).   [208] EPR Sch.2 Part 13.

[209] Department of Housing and Public Works (2013).

[210] If it is a concurrence activity under the EPR Sch.2 and requires development consent, then the Sustainable Planning Regulation 2009 applies. This ensures there is no duplication of regulation under the Integrated Development Approval Scheme; see also Chapter 4.

[211] NWA s.21.   [212] NWA Sch.1.

[213] General Authorisation No.665 of 2013.

much to commend it, with much specificity at the middle tier. Large wastewater treatment plant will require a full licence, which will also incorporate effluent standards. Thus all the jurisdictions take a tiered and proportionate approach.

Wastewater treatment plant are registered with the Department and compliance is assessed and monitored under the 'Green Drop' scheme, paralleling the Blue Drop for drinking water and giving aggregate scores across a set of parameters, including effluent quality and also management and operational processes.[214] Given the known impacts, on groundwater especially, of inadequate sanitation of different types, there is a specific policy protocol on managing this.[215] Management of effluent quality is another priority for DWA.[216]

In every jurisdiction the primary legislation provides for trade effluent (or trade waste) consents, whereby discharges into sewers are controlled by the WSPs, and these rules have many similarities.[217] WSPs must be satisfied about the effect on existing or potential reuse of wastewater or sludge; that the discharge will not harm sewerage, or the health and safety of workers; and that the treatment plant can deal with the effluent. Approval may be subject to conditions, e.g., the quantity, rate, and required pre-treatment; charges are based on formulas reflecting *inter alia* the volume and strength of the effluent. If controls from the environmental regulator on the end-point discharge tighten, so too do the controls on trade effluent and the cost of these licences. The WSP must balance the need to achieve prescribed standards with the need to maintain a revenue stream. If controls tighten too much, one risk is that (if permitted) businesses will install in-house wastewater treatment plant for pre-treatment, and revenues will be lost.

5.4.8   Service standards

In addition to drinking water quality and effluent treatment, there are other service standards that a comprehensive water services law will provide for, such as (especially) water pressure, or sewer flooding; and there may be standards applying to customer service as such.

In England, there are statutory Guaranteed Service Standards.[218] These cover water pressure; response times, to letters and phone calls, or keeping appointments; restoring supplies and if necessary providing emergency supplies; and sewer flooding. Minimal standard payments for breaches are made in most cases, with more substantial payments (refunding the whole annual sewerage charge up to £1000) for sewer flooding, where this is the WSP's liability. In addition WSPs will normally assist with clean-up of sewage, certainly where this is internal. There is also a statutory body to represent the interests of consumers, the Consumer Council for Water.[219] Its predecessor was part of OFWAT, but according to prevailing regulatory theory in the UK is now a separate entity. In addition, Licence Condition G in the conditions of appointment requires undertakers to have a Customer Code of Practice. WSPs may make additional payments beyond the minima specified in the regulations, or compensate for other service failures not covered by the Guaranteed Service Standards. Customer satisfaction has recently been given a higher priority in the business planning and price setting process (below).

In Scotland a similar scheme applies, but is not statutory.[220] As in England and Wales, customer protection was a function of the economic regulator, but again there was reform and, eventually, separation of functions and bodies. Separate Customer Consultation Panels were established, with *inter alia* powers to investigate complaints.[221] However, this body was wound up and its functions transferred to Consumer Futures, a general consumer body.[222] This was in turn affected by UK-wide reforms which have seen consumer advocacy over water services in Scotland transferred to Citizens' Advice Scotland, and the investigation of complaints to the Scottish Public Services Ombudsman.[223] A new, non-statutory, body, the Customer Forum, has recently been established in addition to these statutory bodies, to represent the consumers in price setting.[224] Despite this rapidly changing situation, as in England, there is much more focus on the customer in the current round of economic regulation.

In Queensland, there is a statutory requirement that customers either receive a contract containing details of their levels of service, or receive a copy of the customer service standard (CSS).[225] The CSS will state the level of service to be provided and the process for connections, billing, metering, accounting, consultation, complaints and dispute resolution. There are guidelines from the regulator; the CSS should reflect the standards in the WSP's asset management plan.[226] This will include provision

[214] DWA (2012). The Green Drop programme is perhaps less well developed than the Blue Drop, but an increased number of systems are being assessed.

[215] DWA (2003a).   [216] DWA (2010a).

[217] Sewerage (Scotland) Act 1968 Part II; WIA 1991 Part IV Chapter 3; Water Supply Act Chapter 2 Part 6; SEQ Water Act Chapter 2C; WSA s.7 along with the Standards Regulations and the Model Bylaws (DWAF 2005).

[218] Water Supply and Sewerage Services (Customer Service Standards) Regulations SI 1989/1159 as amended.

[219] WIA ss.27A–27K.   [220] Scottish Water (2012).

[221] Water Services (Scotland) Act 2005 s.3.

[222] Public Services Reform (Scotland) Act 2010 asp.8.

[223] Scottish Public Services Ombudsman Act 2002 asp.11.

[224] 'Customer Forum' see http://customerforum.org.uk/.

[225] Water Supply Act ss.113–120.

[226] DEWS (2010a); see below for economic regulation and asset management.

for water quality, pressure, availability and interruptions, and sewer flooding and overflows.

There are two complaints procedures depending on the nature of the WSP. The Ombudsman Act 2001 provides mechanisms to review 'administrative actions'.[227] If the WSP is an 'agency' to which this applies, including local government, then complaints are made through that route. The Ombudsman can also undertake systematic investigations, and make recommendations. Otherwise, complaints are made to the regulator (DEWS); they have a duty to inquire and may serve a notice, which may require revision of the CSS.[228] In SEQ, the SEQ Water Act also requires provision of a customer service charter by the distributor-retailers,[229] and this must specifically address hardship. The Water and Sewerage Code also applies for households and small businesses, and covers pressure and interruptions to supply.[230] Other service standards such as sewer flooding are expected to be in the service contract.

In South Africa, there is no specific requirement under the WSA for WSPs to provide service standards or customer information. However, the Standards Regulations do address pressure, leakage, greywater and a water audit, which should then be detailed in the Water Services Development Plan.[231] Under the Blue Drop scheme, to achieve the highest scores, there must be engagement with customers, including a customer charter and customer care centre; as well as 'informative billing', and community and schools awareness campaigns.[232] The 2010 Regulatory Strategy also considers customer standards. It includes a set of performance indicators on pressure, interruptions to supply, and sewer flooding; and recognises the need to engage with customers and provide them with opportunities to report breaches.[233] The Regulatory Strategy is comprehensive but there is much that still needs to be implemented and this may depend in part on DWA's success in obtaining the legislative reforms that it has been seeking.

A decade ago, DWAF did report on responses to customers, and on service standards generally, but this did not continue, reflecting the priority of providing basic services and, beyond that, securing acceptable drinking water quality. An incentive-based policy context for improving customer service, with enforcement of regulation where appropriate, seems a good approach, but in addition the Standards Regulations are in need of some revision.

5.4.9 Disconnections and reductions in supply

Disconnections, or reductions in supply, may be necessary because of shortages or emergencies (below), but may also be permitted as a response to non-payment; this may be politically sensitive, especially if PSP is introduced.

In England and Wales, the WIA 1991 did allow disconnections after a court order was obtained, but following a successful judicial review this was prohibited, along with limiting devices, for domestic customers.[234] Non-payment is a political and practical issue for government, the WSPs and the regulators, exacerbated by the recession. Since 1999, there have been regulations protecting metered customers defined as 'vulnerable groups'; these include those on benefits, with larger families or a designated medical condition, who may suffer particularly badly from compulsory metering.[235] However, there are other customer groups affected, and a review recommended explicit provision for wider social tariffs.[236] Water companies are now empowered, but not required, to introduce these, by cross-subsidy from other household customers; government guidance suggests that a levy of 1.5% would be reasonable. If companies adopt a social tariff and it complies with the guidance, then OFWAT should accept it and the WSP may not need to provide under the regulations.[237]

In Scotland, there is no provision to disconnect domestic customers, and no Scottish government has suggested this or is likely to in the prevailing political climate.

In Queensland, water supply can be restricted if a water restriction has been contravened or charges not paid, notice is given, and non-compliance continues.[238] Water may be reduced to the minimum necessary for health and hygiene, but must not be shut off completely. The minimum is not specified. This is one possibility where governments do not wish to allow WSPs to terminate supply, but do wish supply-related sanctions for non-payment; but it might be advisable to specify a minimum amount.

In South Africa, the WSA also allows disconnections for non-payment, but not for basic services where non-payment is because of inability to pay.[239] The provision of FBW may make disconnection less likely, but in early case law, disconnections were upheld where the user consumed more than the FBW entitlement.[240] The model bylaws permit disconnections in only

---

[227] Ombudsman (Qld) Act 2001 No.73.   [228] Water Supply Act s.118.

[229] SEQ Water Act ss.99AD–99AEA.   [230] DEWS (2013).

[231] WSA s.12 requires water services authorities to draw up these plans and s.13 outlines the content.

[232] DWA (undated). Billing, and many other matters relating to service provision by municipalities, are regulated under the MSA 2000. This will be considered below, in business planning.

[233] DWA (2010a) para.6.4.6 and Chapter 16.

[234] WIA 1999 c.9 s.2 inserting new s.69A into WIA; *R v Director of Water Services ex p Lancashire County Council and Others* [1999] EnvLR 114.

[235] Water Industry (Charges) (Vulnerable Groups) Regulations SI 1999/3441 as amended.

[236] Walker (2009); Floods and Water Act 2010 s.44. Confusingly, this does not amend the WIA, which would have seemed sensible.

[237] DEFRA (2012a).

[238] Water Supply Act s.169; SEQ Water Act s.99AT; and DEWS (2013).

[239] WSA s.4.

[240] *Manquele v Durban Transitional Metropolitan Council*, Case No.2036/ 2000. Mrs Manquele did not pay for water used beyond her FBW.

the most limited circumstances, and give further protection to individuals who declare themselves indigent.[241] There are issues around identifying which households are least likely to be able to comply with complex administrative systems.[242] There are also arguments around the level of the FBW entitlement and, linked to that, the actual structure of the block tariffs used, especially the price of the second block. This brings us to the question of economic regulation and business planning.

5.5 ECONOMIC REGULATION AND BUSINESS PLANNING

The outline structure for regulation has been set out already. To a greater or lesser extent, all these jurisdictions accept the need for cost recovery and some element of 'user pays'. Equally, all provide some mechanisms for cross-subsidy (from other service users) and perhaps general subsidy (from general or local taxation). The neoliberal agenda of full cost recovery and unwinding of subsidies has had huge practical consequences, both positive and negative. It has led to a new focus on business planning and efficiency, not just to the benefit of investors, sharpened by the continued failure to meet the needs of the poorest and the global recession.

5.5.1 England

In England, because of divestiture and also because the UK was at the forefront of neoliberalism in the 1990s, there has been a continued focus on the practice of economic regulation, which was instituted for most of the newly divested utilities when they were first sold. The UK decided to adopt and develop a new mechanism of price controls, rather than the more common rate of return regulation, used, e.g., by utilities in the USA.[243] The maximum price increase is set over a period of time, currently five-yearly, during which any efficiencies can be retained by the operator, and then a readjustment is made for the next period.[244] It is intended therefore to incentivise efficiency. Operators may appeal the conditions of appointment, modification, and

determination of prices to the Competition and Markets Authority (CMA),[245] and may ask for an interim determination during the price review period where circumstances change beyond their control.[246] Charging schemes are made annually and approved by OFWAT.[247] The CMA may be asked by OFWAT to investigate whether the carrying out of a company's functions is operating against the public interest and, if so, whether modifications to a company's conditions of appointment would be a remedy. If so, OFWAT has a duty to modify the conditions.[248] The CMA also has a role in decisions regarding the licensed suppliers, which are not examined in detail here. It determines questions under the WIA using a public interest test, rather than a narrower competition law test as such.

The system uses competition by comparison, across the regional monopolies. There must be enough companies operating to provide robust comparators, therefore mergers require a degree of control. OFWAT is one of a group of regulators that has concurrent jurisdiction with the CMA;[249] mergers must be referred to the CMA if the turnover of the company being taken over is more than £10 million/annum.[250] Some special rules apply, requiring the CMA to consider whether the merger will prejudice the ability of OFWAT to make comparisons, weighed against any countervailing customer benefits.[251] Uniquely to the water market, customer benefits will only prevail where they are substantially more important than the prejudice to OFWAT. The CMA also has broad general powers of investigation and sanction.

To support price setting, there has been a very detailed set of business planning requirements for WSPs, enabled by minimal provision in WIA. OFWAT must prepare a forward work programme,[252] report to the Secretary of State,[253] and maintain a register of appointments.[254] The Secretary of State and OFWAT may publish advice;[255] undertakers have a general duty to supply information,[256] and specific duties to report on compensation paid for breaches of service standards.[257] Detailed information

---

entitlement, and the courts approved her disconnection, based on arguments under the WSA. The Council subsequently announced that they would not disconnect her (and others in a similar position) from the FBW, but would use 'trickler' or flow restrictor devices. See, for analysis of the early case law, COHRE (2003).

[241] DWAF (2005); and see Still *et al.* (2007).

[242] COHRE (2008); Algotsson and Murombo (2009).

[243] On the historical process, see Vickers and Yarrow (1988); on price setting and rate of return, see Ogus (1994); on the system introduced in England, see Littlechild (1988).

[244] Determinations and Conditions are provided for under WIA ss.11–12. Price caps are established under Condition B of the Instruments of Appointment, available at http://www.ofwat.gov.uk/industrystructure/licences/.

[245] WIA ss.13–17. The CMA is a new body, established under the Enterprise and Regulatory Reform Act 2013 c.24. It takes the functions of both the Competition Commission and the Office of Fair Trading. As it is not yet fully operational, the WIA at the time of writing refers still to the Competition Commission, but this will be superseded by the time this book is in print. Therefore reference will be made to the new body and its powers.

[246] Licence Condition B; and see generally 'Interim Determinations' http://www.ofwat.gov.uk/pricereview/setting/interim/.

[247] WIA ss.142–150.   [248] WIA s.14.

[249] Competition Act 1998 c.41 s.54, WIA s.31; although there is a new power to remove this by Ministerial Order, Enterprise and Regulatory Reform Act 2013 ss.51–53.

[250] WIA ss.32–35.

[251] WIA Sch4ZA, inserted by Enterprise Act 2002 c.40; further specification is proposed in the Water Bill s.14.

[252] WIA s.192A.   [253] WIA s.192B.   [254] WIA s.195.

[255] WIA s.201.   [256] WIA s.202.   [257] WIA ss.38A, 95A.

is then specified under the terms of the licences; the conditions of appointment are effectively another layer of regulation. Until the current price review period, OFWAT required a comprehensive annual return on the WSPs' business (the 'June return'). However, in recognition of the complexity and regulatory burden of the price setting process, and in order to move to a more 'risk-based' approach, OFWAT removed this reporting requirement.[258] Undertakers now report against key indicators, and make a 'risk and compliance' statement, and a financial return.[259] Key indicators include security of supply, sewer flooding, pollution from wastewater, greenhouse gas emissions and basic financial indicators, as well as a new set of measures addressing customer satisfaction.[260] OFWAT has also rationalised its own reporting on WSP performance.[261]

The annual reporting then feeds through into the periodic review. The WSPs produce draft business plans, and then OFWAT issues a draft determination. Following another round of consultation, a final determination is made.[262] In the 2015–20 price period, there will be separate price controls for wholesale and retail services, intended to increase efficiency but also to enable the market reforms under the Water Bill.

Perhaps the most interesting feature of the new approach by OFWAT is an emphasis on consumers. Currently, concerns are widespread that the industry is more focused on the needs of its investors than its customers. Some WSPs have increased their ratio of debt at the expense of equity, aided by a favourable tax regime;[263] some are paying high returns to their shareholders, leading to significant adverse comment.[264] OFWAT itself has recognised the problem and the concern.[265] The regulatory burden was also seen as part of the problem. Part of the solution to both issues is seen to be in giving a stronger voice to customers in the business planning process, feeding through into price setting. The WSPs were instructed to establish 'customer challenge groups', with representatives from different authorities and regulators, which could challenge the companies on their plans. If, however, the groups approved draft business plans, that would support a 'lighter touch' under the risk-based approach.[266] Results from customer satisfaction surveys are also part of the new set of performance measures.

On the one hand, this seems a very useful approach to reducing regulation and changing the behaviour of the undertakers. Yet it is not wholly clear that the reduction in data will be to the long-term benefit of customers. Although the June return was complex and technical, it did provide data that are not now available, raising issues of transparency and, eventually, accountability. Further, the undertakers remain monopolies, inevitably more inclined to profit maximisation than responding to consumer pressure. The upstream reforms proposed in the Water Bill, and more especially the introduction of retail competition, may make a difference. But, essentially, 'comparative competition' is not competition at all. Renzetti argued a decade ago that it was regulation – of prices, service standards and the environment – rather than competition itself, that improved efficiency in England after divestiture.[267] By that analysis, the early wins were low-hanging fruit; currently, the market is outpacing, and perhaps outmanoeuvring, the regulator. Also a decade ago, Bakker suggested that the water sector was a highly constrained regulatory environment, inherently unattractive to investors, and that a 'mutual' structure would be more appropriate.[268] Yet a stable market with stable returns, not a speculators' market, is what one would expect – indeed hope for – in an essential service. Instead of dividends falling to an unattractive level, they have soared, to great public concern. Perhaps OFWAT did not have, or has not used, the correct regulatory instruments; perhaps, better competition under the next reforms will make a difference; perhaps, consumer input will make that difference; but, perhaps, Renzetti was over-optimistic. Private companies, quite properly, maximise the interests of their shareholders. Perhaps 'the market' is simply too difficult to control.

The next section will explore how a very similar regulatory model has worked in Scotland, in the context of the public sector.

5.5.2   Scotland

In Scotland, the WICS uses the same price cap methodology as in England and Wales, and competition by comparison, requiring SW to report against the same set of performance measures as the English WSPs.[269] As in England, SW can appeal the final

---

[258] 'June Return' see http://www.ofwat.gov.uk/regulating/junereturn/, including links to the previous returns. This step can be seen as part of a general UK drive for 'better' regulation, often defined as synonymous with 'risk-based' or 'proportionate' regulation, and sometimes less regulation. There is no scope in this work to start to unpick these concepts, but it can be seen as a development of the interest in regulatory theory. Proportionality and targeting, along with accountability, are core principles for 'better regulation' in the UK. See also 'Better Regulation' https://www.gov.uk/government/organisations/better-regulation-delivery-office; and OFWAT (2012).

[259] OFWAT (2012).

[260] 'Key Indicators Guidance' see https://www.ofwat.gov.uk/regulating/compliance/reportingperformance/kpi.

[261] See 'OFWAT Publications and Reports' http://www.ofwat.gov.uk/publications#reports.

[262] At the time of writing, spring 2014, WSPs have produced their draft business plans for the 2015–20 price period, and OFWAT will issue the draft determination in the summer.

[263] This was recognised as a problem a decade ago, but no significant changes were made; DEFRA/OFWAT (2003).

[264] Turner/CentreForum (2013), Fortson (2014).

[265] Cox (2013); Johnston Cox is the Chairman of OFWAT.

[266] 'Customer Engagement' see http://www.ofwat.gov.uk/pricereview/pr14/customer/.

[267] Renzetti and Dupont 'Ownership and Performance of Water Utilities' in Chenoweth and Bird (2006).

[268] Bakker (2003a).   [269] WISA s.47, s.56.

determination to the CMA.[270] In England now, though, these measures are no longer used to assess overall performance by OFWAT; customer satisfaction is being given much more prominence. In Scotland, it is intended to retain the technical performance measures at least through the next price review, as they will still indicate relative performance improvements by SW. Customer satisfaction will also be measured, but as SW does not bill customers directly, and in England most customer interaction is around billing, it would not be possible to replicate the new English system directly. In Scotland, as in England, the regulatory process has evolved and less detail has been required, both from SW and from WICS.[271]

Strategic reviews of charges took place in 2002, 2006 and 2010; the next determination is currently being negotiated to start in 2015 when the period will extend again, to six years. This will align with the planning period for the RBMPs. Along with a formalised system of interim allowances at three-year intervals, this is also expected to allow a much smoother capital programme.[272]

The core functions of SW are funded by fees and charges and by borrowing from Government; it does not receive grant aid.[273] Currently the Government receives payment of interest and capital but does not take a 'dividend'; and SW is enabled to retain outperformance on its regulatory settlement during the price period as an incentive, just as would happen in England. It has a number of subsidiaries which are fully 'ring-fenced' from the core, enabling it to participate in joint ventures for the capital programme, engage in outreach and training internationally, and develop projects such as energy generation or waste management that should not be funded by core customers or expose them to risk. SW Business Stream, providing retail services to businesses and competing with the licensed providers, is the most important.[274] Water charges for domestic users are collected by local councils along with local taxation. This has an administrative benefit for SW, and also allows charges to be banded reflective of property values, which gives some protection to the poor, but it does mean that domestic customers have no direct link with their supplier.

The policy framework is set by Government via a Statement of Objectives,[275] and then the Principles of Charging.[276] The first sets out the essential and desirable elements of the investment programme and service operation, and the second establishes principles, including, currently, that charges should be cost-reflective, stable (rising by no more than inflation), and recover the full costs of the service. Although there has been some unwinding of cross-subsidy in recent years, for domestic charges there is still harmonisation within charging bands, to smooth the costs of delivery in rural areas. Most cross-subsidy within the business sector has been unwound. The Government has consulted on both the Objectives and the Principles for the next price period,[277] and SW has produced a draft Business Plan,[278] along with a longer-term Strategic Projection over 25 years.[279] Again, this longer-term view is intended to signal a move away from very rigid pricing and especially investment periods, with significant peaks and troughs that affect the supply chain. It is also reflective of a maturing regulatory environment. In the earlier price reviews, there was an adversarial tenor to the relationship between SW and both WICS and to an extent the Scottish Government, and the draft business plan and draft determination were some distance away in terms of both prices and service standards. This may be a normal negotiating tactic, but it did not evidence a consensual approach. That consensual approach has been much more apparent in the current negotiations, and both complemented and encouraged by a different approach from the WICS and a more important role for consumers. The WICS, along with Consumer Futures and SW, and supported by the Scottish Government, established the new Customer Forum, and empowered it to negotiate with SW over the discretionary elements of the spending programme.[280] The WICS then gave the Forum and SW to understand that if they agreed a settlement that was reasonable, within certain parameters, WICS would agree to the draft business plan.[281] The intention was both to empower customers and to create more 'ownership' of the settlement by SW's Board. Although the process is still underway at the time of writing, it seems to have been very successful.[282]

Although on the face of it this has many similarities to the English customer challenge groups, set up for very similar

---

[270] Water Services etc. (Scotland) Act 2005 (Consequential Provisions and Modifications) Order SI 2005/3172.

[271] For the voluminous documentation on the previous price reviews, and the reports produced in previous years on different aspects of performance, see 'WICS Publications' http://www.watercommission.co.uk/view_Publications_Main.aspx.

[272] For the current thinking behind the regulatory system in Scotland, see WICS (2013).

[273] The core functions were recently further clarified; Water Resources (Scotland) Act 2013 s.26.

[274] For discussion of the various activities that SW can undertake through its subsidiaries, in the context of the Scottish Government's aspiration to build a 'Hydro Nation', see Scottish Government (2010); Scottish Government (2012). This aspiration resulted, inter alia, in the Water Resources (Scotland) Act 2013.

[275] Currently, Scottish Water (Objectives for 2010–2015) Directions 2009; this is a Direction under WISA s.56–56A.

[276] Principles of Charging for Water Services 2010–2015; this is a policy statement under WISA s.29D.

[277] Scottish Government (2012a).   [278] Scottish Water (2013a).

[279] Scottish Water (2013).

[280] The mandatory elements – meeting statutory objectives such as drinking water quality – are of course not negotiable.

[281] See WICS (2013).

[282] I must declare an interest, as one of the members of the original Forum appointed in 2011. The insights gained were invaluable; I and my colleagues sincerely hope, and genuinely believe, that our role has been of benefit.

reasons, there are differences both in structure and in result. The Forum, unlike the challenge groups, is not composed of regulators and public authorities. On the other hand, the Forum has had significant input from all SW's regulators, and from the Scottish Government. In England, the process is still essentially adversarial; in Scotland, it has shifted. This can only partly be placed at the feet of the Forum. Perhaps much more important is a cohesive understanding of the goals and objectives, and ethos, of the public service. After a number of reviews when the WICS obtained and scrutinised large volumes of data, and challenged SW's performance in many respects, it now seems that all the parties are working together, to provide the best service at the best price. That is surely a success story.

This may be due to relatively clear governance arrangements – the legislation in Scotland is not perfect, but it is relatively coherent and understandable. The accountability and reporting lines are clear, as are the roles of the various institutions, and the policy objectives. Yet the general policy and governance arrangements are very similar to those in England, where the current price review is as acrimonious as any. A decade ago, SW performed at the lowest range, and sometimes significantly below the lowest, of the English WSPs. It now performs at the upper quartile. Scotland followed a similar path to many countries in removing water services from local control. It was unusual in establishing an independent regulator to set prices whilst maintaining a public sector organisation. It does seem that the Scottish model has succeeded in regulating the public sector to private sector efficiency, whilst still maintaining a public sector ethos that is not diverted by the needs of investors or shareholders.

5.5.3 Queensland

In Queensland, the COAG reforms promoted better economic regulation, cost recovery and transparency, including corporatisation or commercialisation of public water services.[283] Given the multiplicity of providers in Queensland, this section will focus on the obligations under the Water Supply Act and the SEQ Water Act, with some reference to the regulation of bulk services, and to local government as such, but it will not look in detail at the regulation or governance of the water authorities,[284] at the regulation of bulk supply for irrigation by SunWater,[285] or at the 'corporate governance' of the distributor-retailers.[286]

Government controls the price of bulk water, supplied mainly by SEQWater and whether or not the water is supplied in SEQ. The QWA provides for a Bulk Water Supply Code and bulk water supply agreements, and the Minister can create mandatory terms and has powers of direction.[287] The Minister may seek advice (generally, though it is not specified, from the Queensland Competition Authority, QCA), but the decision is binding and is expressly not subject to any form of review or appeal except in the event of a successful judicial review on grounds of error.[288] The Code itself is also binding. As well as pricing and access to networks, it includes operating protocols for infrastructure, bulk water quality, metering, and transitional arrangements.[289] The QCA may be nominated as an Investigating Authority under the Code and may be asked to consider the bulk price and service charges, and other matters as the Minister directs, relevant to pricing or access. Draft reports on pricing will be published for comment, but not reports on network access; the Minister is not bound by the QCA's reports. So, unlike Scotland, that is a more typical situation for public suppliers, where government takes (or may take) advice on pricing, but the price is not set by the external agency.

These provisions were part of the rationalisation of the bulk services. Under the reforms of 2007–12, a 'price path' was established to impose the full costs of bulk supply on all SEQ providers by 2018. As a result of the more recent rationalisation, some reduction in these increases has been mandated, to be passed on to the consumers via their WSPs.[290]

Also under the QWA, SEQWater and designated providers must have a water security programme, with specified levels of service.[291] The service levels will be subject to consultation. The water security programme will include infrastructure planning, demand management and drought, and will be submitted in draft to the Minister; any designated provider will not then need a Drought Management Plan under the Water Supply Act. The water security programme will replace the current regional SEQ Water Strategy.[292] There are also regional strategies for other parts of the state, which will not be replaced.[293]

The Water Supply Act contains planning and regulatory matters, variously applying to all WSPs or only to those outwith SEQ; the SEQ Water Act has some different provisions, whilst

---

283 COAG (1994, 1995, 2004).   284 QWA Chapter 4.

285 Under the QWA, but also the Government Owned Corporations (Qld) Act 1993 No.38. This provides for corporatisation, as well as financial accountability and other elements of corporate governance.

286 The SEQ Water Act provides a complex regime for accountability and decision-making for the distributor-retailers, and their relationships with their constituent councils. For a thorough and critical analysis of these rules, see Baumfield (2012).

287 QWA Chapter 2A Part 2; DEWS (2012a). Some provisions apply to all bulk customers, and some only within SEQ.

288 QWA s.360Y. This is an unusual provision in its extent and specificity.

289 From the previous entities, and the system of Market Rules under the Grid; see DEWS (2012a) Chapter 7, and Water (Transitional) Regulation 2012 SL No.242.

290 In SEQ bulk costs will be displayed on customers' bills. See generally 'Bulk Water Prices' http://www.dews.qld.gov.au/policies-initiatives/water-sector-reform/water-pricing/bulk-water-prices.

291 QWA Chapter 2A Parts 1–2.

292 Queensland Water Commission (2010).

293 Queensland Government (2006, 2010, 2012).

the Local Government Act is also relevant. All service providers must be registered with DEWS,[294] and there is a process for transfer (for example, if a private entity, or a water authority, stops or starts serving an area). There is specific provision that registration does not imply an entitlement to water,[295] and also that ownership of infrastructure will not pass to the owner of the land.[296]

Service providers have a series of planning obligations, and it is through these that their activities are publicised, and approved where necessary, by the regulator (the chief executive).[297] Strategic Asset Management Plans (SAMPs) must be certified by an engineer and then approved unless they are 'inadequate in a material particular'.[298] This applies to all WSPs except small rural providers. System Leakage Management Plans are also certified and approved in the same way, but WSPs can apply for an exemption. A series of grounds are given, including that the system is relatively new and water efficient; that it is operating as a groundwater recharge scheme; or that it would not be cost-effective.[299] Drinking Water Quality Management Plans (DWQMPs) are required, and the regulator may seek advice before approving these (for example, from Queensland Health); there is an obligation to report any non-compliance. Customer service standards have been discussed above. If a WSP obtains bulk supply from infrastructure not covered by a DWQMP, it may request relevant information from the operator. There are specific enforcement procedures for breaches of drinking water quality requirements.[300]

Plans must be regularly reviewed and audit reports provided; the regulator can determine periods, and for SAMPs there is specific provision for review to meet new industry best practice. The regulator has wide powers of direction, can carry out spot audits, require information, and issue show cause notices. Service providers must report annually on their plans and compliance, unless they are local governments reporting under the Local Government Act and a copy of that report is passed to the regulator.[301] There is extensive guidance available through DEWS to assist WSPs with preparing plans and reporting on their obligations.[302] Drought Management Plans are also required (below).

For the distributor-retailers, under the SEQ Water Act, separate integrated 'Water Netserv Plans' may be prepared, and when this is done they will replace SAMPs, System Leakage Management Plans, and Drought Management Plans.[303] These should be in place by March 2014, and reviewed every five years; there is

some detail in the Act. They will cover *inter alia* infrastructure, service standards, future demand, leakage and sewer overflows; they must have regard to relevant SEQ regional strategy documents.[304] The SEQ Water Act also provides for infrastructure work, such as breaking roads, both for the WSPs and for other public entities whose activities may affect water infrastructure;[305] the SEQ WSPs jointly are required to prepare a Design and Construction Code.[306]

For local government WSPs, as noted above, the Local Government Acts provide for 'competitive neutrality' where there is a 'significant business activity'.[307] If turnover is above the threshold, local governments must consider two-part tariffs.[308] Ministers may order WSPs to provide services non-commercially, under a Community Service Obligation, which must be accounted for.[309] Complaints over competitive neutrality are made to the QCA, which as well as being the Investigating Authority for bulk services, has a general jurisdiction over water pricing.

'Government agencies', which would include local governments, water authorities, and SunWater as a Government Owned Corporation, are controlled under the general competition rules.[310] The QCA may ask the Minister to declare a 'monopoly business activity' even where it would not otherwise be 'significant',[311] and this has been done for several local governments and the distributor-retailers.[312] The Minister may then refer their pricing practices for investigation; any subsequent recommendations must be kept on a public register.[313] The QCA hears complaints over competitive neutrality.[314] It has jurisdiction over access to infrastructure, including water and sewerage, and can approve access undertakings, or make access determinations if the service has been declared for this purpose.[315] The QCA developed a set of 'better regulation' principles for water pricing: that prices should be cost-reflective; be forward-looking; ensure revenue adequacy; promote sustainable investment; ensure regulatory efficiency; and take account of the public interest.[316] It is

---

294 Water Supply Act s.20.   295 Water Supply Act s.29.
296 Water Supply Act s.30A.
297 Water Supply Act Chapter 2 Part 4; the regulator is identified in s.10.
298 Water Supply Act s.74.   299 Water Supply Act s.84.
300 Water Supply Act Chapter 5 Part 5.   301 Water Supply Act s.141.
302 See especially DEWS (2010b).   303 SEQ Water Act s.53AL.

304 SEQ Water Act Chapter 4B. Relevant documents would include the regional water supply strategy, Queensland Water Commission (2010), and also, e.g., strategic planning frameworks.
305 SEQ Water Act Chapter 2B.   306 SEQ Water Act Chapter 4A.
307 Local Government Act ss.43–48.
308 Local Government Regulation Reg.41. In SEQ, this is required for the distributor-retailers and the tariff is further divided to also show the bulk charge paid by the distributor-retailer.
309 Community Service Obligations are provided for under the Local Government Regulation Regs.24, 36, 41; the Government Owned Corporations Act 1993 ss.112–113 (for SunWater); and the QWA s.582 and s.683 (for Category 1 Water Authorities).
310 Queensland Competition Authority (Qld) Act 1997 No.25 (QCA Act) Part 3.
311 QCA Act s.20.
312 Queensland Competition Regulation 2007 SL No.207, Reg.2A.
313 QCA Act s.30A.
314 QCA Act s.42 ff; the QCA may investigate and report.
315 QCA Act Part 5.   316 QCA (2000).

currently developing a long-term pricing framework for SEQ,[317] but currently, though the QCA monitors retail prices, it does not determine them.

The QCA also has specific powers to set prices for water services by suppliers that are not Government agencies, and must do so for any declared monopoly supply activities.[318] There are criteria for this, including business efficiency, competition and the interests of consumers.[319] In many ways these are similar to the factors to be considered by OFWAT and the Secretary of State in England, but they would apply only to any private providers. Although Queensland does not prohibit these, most supply is from some category of public authority. Some rural Water Authorities may be constituted as private firms or cooperatives, but for all regulatory purposes are treated as public bodies.

5.5.4   South Africa

In South Africa, like Queensland, the focus of this section will be on distribution and retail services, mainly through local government, with a brief discussion of the control of bulk supply through the Water Boards. Central government is responsible for economic regulation of water services, through DWA, but municipalities report to CoGTA, and to the Auditor-General.[320] DWA has an ongoing project looking at pricing (of raw water), tariffs (for the whole value chain), and economic regulation, including potentially an independent regulator.[321] The 2010 Regulatory Strategy evidences the need for a comprehensive approach, but it is likely that further progress will depend on political negotiations around responsibility for sanitation, appropriate modes of regulation, and the current wish of the Department to bring together the NWA and WSA.

For bulk supply, the Water Boards report to the Minister, and prepare five-yearly policy statements and yearly business plans.[322] Effectively, therefore, the Minister approves the prices set for raw water. The Minister may direct amendments, and may issue directives to the Boards. The 2010 strategy recognises the need for a clear separation between the Department's role as shareholder, and the regulatory function.

For distribution and retail services, there are regulations on tariffs.[323] When drawing them up, the Minister must consider *inter alia* social equity; financial sustainability; cost recovery; a return on capital; and the need to provide for drought and 'excess availability'.[324] The regulations provide for cost recovery,

overheads and maintenance, and the cost of capital. They require providers to differentiate communal supply or supply at controlled volumes; and for sanitation, whether or not there is a sewered system; to use a block tariff with the lowest block set at the lowest viable amount; and to prioritise basic services.

There are also specific regulations on contracts for water services.[325] These should again ensure that the service is 'efficient, equitable, cost-effective and sustainable'.[326] The regulations then provide some structure on *inter alia* the scope of the contract; performance targets and indicators; duration of no more than 30 years; monitoring; accounts, annual reports and access to information; and a consumer charter.

So the essence of a control regime is provided by DWA, and the core principles for economic regulation reflect those in every regime studied here. However, as almost all services are provided by municipalities, it is also necessary to look at the broader structures for delivery of municipal services.

The Municipal Systems Act (MSA) sets the general framework. It establishes general powers and requires community participation, development planning and performance management, as well as providing for administration and services. Municipalities must produce Integrated Development Plans[327] and the Water Services Development Plans may be incorporated into these.[328] There is a general duty to provide basic services, and these should make 'prudent use' of resources and be financially and environmentally sustainable.[329] There must be a tariff policy, supported by bylaws; and these must be compliant with the Municipal Finance Management Act.[330] Tariffs should be equitable, based on use, cost reflective, provide basic services, facilitate financial sustainability, and may differentiate users as long as this is not discriminatory.[331]

A variety of mechanisms may be used for municipal service delivery.[332] Municipalities may provide services internally, or externally, which may be via another public body (including national, provincial, municipal governments, traditional authorities or NGOs) or a 'municipal utility'. If external, there will be a service delivery agreement; if there is an external provider for basic services, there must be consultation. The municipality must assess the best mode of provision, and if considering an external provider, assess the costs and benefits and consult with the community. If there is an external provider, the municipality is responsible for regulation and monitoring, and must ensure continuity of service; the agreement must provide for dispute

[317] QCA (2013).   [318] QCA Part 5A.   [319] QCA Act s.170ZI.
[320] See Auditor-General (2012).
[321] 'Pricing and Economic Regulation Reforms Project' see http://www.
    dwaf.gov.za/Projects/PERR/Default.aspx.
[322] WSA ss.39–40.
[323] Norms and Tariffs in Respect of Water Services Regulation No.652 of
    2001 (Tariffs Regulations).
[324] WSA s.10.

[325] Water Services Provider Contract Regulations No.980 of 2002.
[326] WSA s.19.   [327] MSA Chapter 5.   [328] WSA s.12.
[329] MSA s.73.
[330] MSA ss.74–75; Municipal Finance Management Act 2003 No.56. The
    latter provides governance and accounting systems for municipalities. It
    also regulates PPP schemes, requiring them to be value for money, and if
    providing a 'municipal service', the MSA should apply.
[331] MSA s.74.   [332] MSA Chapter 8.

resolution. There are rules on competitive tendering, which do not apply where the agreement is with an organ of state or another municipality (but in the latter case, there must be a feasibility study). The Minister has broad powers to issue guidance and regulations on tariffs.

There is also provision on 'municipal entities'.[333] These may be private companies, 'service utilities' or 'multi-jurisdictional service utilities', but, if a private company, control must rest with a municipality, or another organ of state. It will then be treated as a public entity for financial management,[334] and cannot have any functions outwith the competence of the municipality. In 2011/12 there were 60 municipal entities, and 278 municipalities.[335] Municipalities may establish 'service utilities' if this would be the best way of providing a service. They may also establish 'multi-jurisdictional service utilities', and the Minister may request this. All of these structures can be used to deliver a water service, within or across municipalities and either alone or in combination with other services.

Institutions may use any source of funds to subsidise a water tariff, and must consider the right to basic services when deciding what to subsidise;[336] cross-subsidy is encouraged in the Strategic Framework. Funding for basic services and capacity-building is specifically envisaged in the MSA.[337] Grants are provided by central government: the Municipal Infrastructure Grant for basic services, and the Equitable Share. The Municipal Infrastructure Grant was intended to end by 2013, on the basis that service backlogs would be completed, and is allocated for services as detailed in the Integrated Development Plan. The Equitable Share is meant to contribute to operational costs and is not ring-fenced, so there is no obligation to spend it on water or any particular service. Under the current Revenue Act,[338] as well as the Municipal Infrastructure Grant, there is a Human Settlements Development Grant and an Urban Settlements Development Grant, for the Provinces;[339] for municipalities, a Municipal Water Infrastructure Grant and Water Services Operational Subsidy Grant, administered through DWA, and a Rural Household Infrastructure Grant, through DHS;[340] and a Water Services Operational Subsidy Grant and Regional Bulk Infrastructure Grant, both under DWA, for designated programmes.[341] Thus there is significant general subsidy, as would be expected where there is great unmet need and political will. DWA recognises the need for financial ring-fencing of water services,[342] but this is not fully achieved.

The Tariffs Regulations provide for block tariffs, with at least three blocks, which may set a free allowance for the lowest block, or may set that at a 'lifeline' level and then make provision for FBW, usually to the indigent poor. The structure of the tariffs is linked to the rates for the blocks; it is expected that high-volume users will pay more, and cross-subsidise the poor. The rate of the middle block is critical. If it is too expensive, especially if the lowest block is at a low volume, then many poor people will be disadvantaged if they move into the second block. In some authorities, prepayment meters may worsen this and the poor may pay more for the middle block than better-off users without meters.[343] If the FBW allowance was raised, then that would be less of a problem. There are particular problems in very small rural authorities and those with no, or very few, high-volume users for cross-subsidy.

Some critics suggest targeting high volume users for non-payment, or allocating FBW on a geographical basis rather than using complex rules on indigency.[344] Others argue *inter alia* that the adoption of commercial principles makes them likely to adopt harsh debt recovery mechanisms and to consider disconnections, even compared to the private sector.[345] This reflects the polarity of the debate. Yet only by clear financial accounting will it ever be possible to truly assess the cost or the value of the asset base, in order to achieve a more effective service. A commercial approach to costs and assets is compatible with a pro-poor policy, given an appropriate policy environment. The difficulties with delivering services through multiple municipalities of very different sizes and resources are encountered in many places, and have led to structural reforms, including in Queensland and Scotland. It is not clear that this structure is working for South Africa, for water or for sanitation. A more focused institutional structure might be better, but location of the service within local government seems unlikely to change.

5.6 WATER CONSERVATION

Water conservation seems an appropriate place to conclude the substance of this book. It is central to the global policy agendas, and to the broad strategic issue of resource management, with which the book began. Across the world, water managers seek ways of conserving existing resources, by reducing demand and increasing efficiency of use, as an alternative to increasing supply. This imperative is linked to water security, heightened in times of drought, and affected by every aspect of environmental change. This final section will consider drought and emergency planning, metering and leakage, use of greywater,

[333] MSA Chapter 8A, inserted by the MSA 2003.
[334] MSA s.86D; Public Finance Management Acts 1999 No.1, No.29.
[335] Auditor-General (2012).   [336] Tariffs Regulations Reg.3.
[337] MSA s.10A.   [338] Distribution of Revenue Act 2013 No.2 (DoRA).
[339] DoRA Sch.5A.   [340] DoRA Sch.5B.   [341] DoRA Sch.6B.
[342] DWA (2010).

[343] COHRE (2008).   [344] COHRE (2008).
[345] See Smith 'The Murky Waters of Second Wave Neoliberalism: Corporatisation as a Service Delivery Model in Cape Town' in McDonald and Ruiters (2005).

recycled wastewater, water efficiency, and the management of stormwater and urban runoff.

In every jurisdiction there are relevant high level duties and/or principles. In South Africa, the WSA has an objective of promoting 'effective water resource management and conservation'.[346] In Queensland, the purposes of the QWA are sustainable management and efficient use;[347] for the Water Supply Act, to provide 'safety and reliability' of supply;[348] in SEQ, efficient service and better management.[349] In Scotland, there are duties on Ministers and SW to 'promote the conservation and effective use' and the provision of 'adequate water supplies'.[350] In England, there are duties in the WIA around conservation and environment,[351] and a specific duty to 'promote efficient use'; OFWAT can require actions to achieve this.[352]

5.6.1 Drought and emergency

In England and Queensland especially, there are extensive provisions on drought, but there are also powers to limit supply in other emergencies. In South Africa, there is a general power to limit or discontinue supply, and in emergencies no notice need be given.[353] WSPs must still 'take reasonable steps' to provide basic services.[354] In Queensland, there are powers for Ministers to declare 'water supply emergencies' arising out of drought, contamination or infrastructure failure.[355] In Scotland and England, there is provision for emergency supplies, e.g., by tanker or standpipe, and involving local government, where water being supplied is 'insufficient or unwholesome'.[356]

In England, WSPs must prepare statutory drought plans,[357] within a policy context set by the EA and the Government.[358] These plans reflect the priority given to domestic supply and the avoidance of interruptions to supply, and are indicative of the variable resource availability. The Drought Plan of Anglian Water, where water is scarce, is very different from that of United Utilities, where restrictions on usage are rare.

The Plans are intended to avoid the need for drought orders or drought permits. Drought orders are granted by the Secretary of State on application of either the EA or the undertakers, and can contain wide powers, to permit, restrict or vary abstraction or discharge of water, either by the undertakers or by others.[359] Ordinary drought orders apply where there is a 'serious deficiency' of supply, or a deficiency 'affecting flora and fauna'. For emergency drought orders, there must be a serious deficiency that may also affect economic and social wellbeing; these may authorise the use of tanks or standpipes. They last for three or six months respectively. Drought permits are granted by the EA on application of the undertaker, and last for a year.[360] These allow the undertaker to abstract water and can modify the terms of their abstraction consents. Undertakers have general powers to institute temporary bans to respond to water shortages, such as the use of hosepipes;[361] and the permissible categories for restrictions can be extended under a drought direction issued by the Secretary of State.

In Queensland, the QWA already allowed restrictions on rights to take water, including for stock and domestic use, where there was a shortage or due to harmful substances.[362] Extensive new powers were introduced during the drought.[363] The Minister can declare a water supply emergency, by declaration, or by regulation for up to one year, because of drought, contamination or infrastructure failure. There must be a 'demonstrably serious' risk of not meeting 'essential' water supply needs. The declaration or regulation will then contain measures to be applied by service providers, which may involve rights to take water, prohibitions on taking water, use of infrastructure and compensation for the same, as well as restrictions on use of water.

Also in the QWA, as discussed above, is a requirement for SEQWater, and nominated service providers, to have a water security programme, including service levels;[364] if this is in place then nominated providers will not require a Drought Management Plan.[365] These are required from most other authorities, except for small rural WSPs and those providing only drainage services or supplying to those holding a water entitlement;[366] there is a further exemption if 70% of the water comes from desalination or the Great Artesian Basin.[367] Drought Management Plans must be consistent with any ROP, and consulted upon with holders of ROLs or Interim ROLs (Chapter 3). The regulator may also require a WSP to have an Outdoor Water Use Conservation Plan, to encourage efficient use by customers, if there is a risk to water security and sufficient measures are not in place.[368] For SEQ, Water Netserv Plans will incorporate Drought Management Plans.[369] There are general powers to restrict supply, including the volume, duration or types of use, on a broad set of grounds, by notice or if directed by the regulator.[370] Directions may be issued if there is a 'significant threat' to 'sustainable and secure supply' outwith SEQ.[371]

---

[346] WSA s.2.   [347] QWA s.10.   [348] Water Supply Act s.3.
[349] SEQ Water Act s.3.   [350] Water (Scotland) Act 1980 s.1.
[351] WIA ss.3–5. These are focused on recreational use and access, and apply to the Secretary of State and the Authority as well as the undertakers, as is necessary where the provider is a private entity.
[352] WIA ss.93A–93B.   [353] WSA s.4(3).   [354] WSA s.11(5).
[355] QWA ss.25A–25R.   [356] Water (Scotland) Act s.76D; WIA s.79.
[357] WIA ss.39B–39C; Drought Plans Regulations SI 2005/1905.
[358] DEFRA (2011); EA (2011b).   [359] WRA 1991 ss.73–79.

[360] WRA s.79A.
[361] WIA s.76, Floods and Water Act 2010 s.36, Water Use (Temporary Bans) Order SI 2010/2231.
[362] QWA ss.22–25.   [363] QWA ss.25A-25ZE.
[364] QWA Chapter 2A Part 2.   [365] QWA s.360B.
[366] Water Supply Act s.122.   [367] Water Supply Act s.126.
[368] Water Supply Act s.133.   [369] SEQ Water Act s.53AL.
[370] Water Supply Act s.41.   [371] Water Supply Act s.42.

In Scotland, despite the abundant resource, some small rural supply zones can suffer scarcity during drought, or as a consequence of infrastructure failure. A broadly similar regime to England exists, with ordinary and emergency 'water shortage orders', and a rather broader set of uses that can be restricted, including more commercial uses.[372] There is nothing directly comparable in South Africa, although the general powers are broad, and there is policy on conservation and demand management (below).[373] The model bylaws do make some provision for emergencies.

5.6.2 Metering and leakage

Metering is arguably a mechanism and incentive for customers to reduce both consumption and bills. This reflects principles that the polluter or user should pay, but disregards the problems of large families, or those with medical conditions requiring large volumes of water, who may benefit from a non-metered charge. If there is a two-part tariff with a fixed infrastructure charge and a volume charge, and the former is relatively high, then there may be no incentive.

Installation, maintenance and reading of water meters is expensive. Reliable meters for supply zones are essential, but meters will not always be justified at household level if the resource is abundant. In England, there has been a steady increase in the number of metered domestic users, but there are still many non-metered households. Generally, householders may choose to have a meter installed.[374] If undertakers declare 'water scarce area status', they may institute compulsory metering, as part of a package of demand management measures; the Secretary of State may determine these areas, but will not direct metering as such.[375] In Scotland, there is metering of commercial users, but virtually no domestic metering, though householders may request a meter if they think this will benefit them. Metering is of public concern, although there is likely to be continued pressure from the EU to introduce this as part of a drive for water pricing and water efficiency under the WFD.[376] In South Africa, as in Queensland and many other countries, there is a presumption that all customers, domestic and commercial, should be metered.[377]

Leakage from networks is complex; it is not possible to reach zero levels. Changes to pressure or to volumes consumed will affect apparent leakage, and greater efficiency in customer use may result in higher proportional leakage rates. Regulators may assess the economic level of leakage (ELL), the level beyond which it will be more expensive to reduce leakage than to increase supply. Such calculations will depend on the wider resource management context and the availability of further resources; the concept of the 'Sustainable ELL' is being used to recognise this.

In England, leakage reduction has formed part of the regulatory regime since divestiture. Undertakers are specifically required to assess their ELL by a cost–benefit analysis weighing up the cost of leakage reduction against the cost of providing a replacement supply. In Scotland, leakage targets have been set since 2008. In Queensland, WSPs must have System Leakage Management Plans, or in SEQ, include this within Water Netserv Plans. In South Africa the 2010 Regulatory Strategy recommends performance indicators for the volume of unaccounted-for water, which includes leakage.[378]

5.6.3 Water reuse and recycling

The reuse of wastewater is a matter of global interest.[379] This might include reusing greywater (e.g., from baths and washing machines, as distinct from 'black' sewer water) within a household; on a larger scale, reusing treated wastewater from various sources, including for irrigation with appropriate public health protections; replacing potable water where this is not required; managing stormwater and surface water runoff; and instituting water conservation measures such as water efficient buildings and appliances. There may be public health concerns that make regulators reluctant to extend its use; these measures require coordination with other areas of regulation, such as planning and building control, and Australia has been at the forefront here, so this section will begin with Queensland.

Queensland is the only jurisdiction to make detailed provision for the reuse of treated wastewater, including its return into the drinking water supply, as a result of the drought, and this remains in force despite the recent rationalisation.[380] If recycled water is to be supplied, there must be a Recycled Water Management Plan, including a risk and hazard assessment, approved by the regulator. If it is returned to a drinking water supply, this must also be addressed in the relevant DWQMP, and there must be a validation programme. Providers may request an exemption, but not if there is a return to a drinking water supply or if recycled water is being supplied to premises under a dual reticulation system. In both these cases there must also be public reporting on the scheme, and this is also the case for the supply of recycled

---

[372] Water Resources (Scotland) Act 2013 ss.38–51, Schs.1–2.

[373] DWAF (2004c); DWAF (2004d).   [374] WIA s.144A.

[375] Water Industry (Prescribed Conditions) Regulations SI 1999/3422, SI 2007/2457; EA (2013d) (the advice on which the Government will make its determination).

[376] European Commission (2012c). Ironically, the EU accepts that generally water services do recover their costs in the UK; across Europe as a whole it is the agricultural sector that pays least for its water.

[377] Standards Regulations, Reg.13.

[378] DWA (2010a) para.6.3.3.

[379] WHO (2006); European Commission (2012e) Chapter 6.

[380] Water Supply Act Chapter 3.

water from coal seam gas operations, to which these provisions also apply, with some variations. Here there may be post-supply conditions to ensure no subsequent contamination of the aquifer. Recycled water schemes may be declared 'critical' if they provide essential continuity of supply, where some additional requirements are imposed. Specific to coal seam gas aquifer recharge, these schemes may be exempt if they have 'no material impact' on drinking water supply.[381] This will be based on the hydraulic connectivity between the aquifer and any source of drinking water, as well as the volume of discharge.

The Water Supply Act has powers to require non-residential customers to draw up water efficiency plans for their operations.[382] In England and Scotland, large industrial and commercial users are at the forefront of water efficiency gains, and in Scotland, this is seen as the area in which licensed retail providers can add value to the basic service, but it is unusual to mandate this.

Under the last set of reforms, there was a mandatory requirement for new homes, and commercial premises, to install water-saving measures such as rainwater tanks or greywater systems.[383] This is no longer required, but the Building Code provides standards where such systems are being used.[384] The Plumbing Code authorises small-scale greywater systems as well as onsite sewerage.[385] There is national and Commonwealth legislation for the registration of water efficient products.[386]

In England, OFWAT looks to see if companies have effective pricing policies, education programmes properly directed to appropriate customers, and an economic level of company activity depending on the location, as pressure is so much greater in the southeast. As with the ELL, OFWAT's main concern has been that undertakers should not spend more on promoting efficiency than it would cost to supply the amount of water saved. However, as abstraction charges rise, or new developments increase consumption in water-stressed areas, or climate change affects rainfall, the wider picture within which the regulator is neutral will also change.

Despite the pressure on resources in the south, and the general impact of climate change and population growth, the UK government has been reluctant to impose the best available technologies or practices for water saving, even on newbuild, preferring the deregulatory approach now found in Queensland. The Sustainable Housing Code establishes relevant standards, but is

not mandatory.[387] Building standards are mandatory where applicable and do provide *inter alia* for greywater use, rainwater systems, composting toilets, and use of non-potable water in some circumstances, as well as effluent from septic tanks and similar systems.[388] Low-flush toilets and other plumbing standards are also required via the WSPs.[389] Shower heads, like washing machines and dishwashers, are not regulated in the UK, though there are labelling schemes. This, along with standards for reuse of water, and better labelling, are being considered by the EU.[390]

In Scotland, there has been less emphasis in the past on water reuse and efficiency, but again Building Standards apply;[391] greywater systems are unusual but there is now provision for these. The equivalent of the Water Fittings Regulations are made through bylaws.[392] SW does carry out education campaigns, but currently these are mainly focused on discharge of inappropriate items to the sewers.[393] Larger businesses, who have switched to licensed retail providers, are likely to have done so in order to obtain tailored water efficiency advice. Concerns over long-term security of supply, especially in the east of Scotland, are reflected in the current long-term strategic projection.[394]

In South Africa, the Standards Regulations make some provision for water fittings and conservation, including the use of greywater and effluent. As well as the requirement to meter premises, the regulations expect water services authorities to conduct water and effluent balance analyses to assist with measuring leakage. The model bylaws provide, for example, detail on metering; the power to require customers to undertake water audits; approvals for installations; constraints on tap and shower flows; and limitations on cistern size.[395] Both the regulations and the strategies make reference to the South African Building Standards, for example for water supply and drainage, and guidance on these standards is available.[396]

Water Conservation and Water Demand Management Strategies were produced in 2004, including a general strategy and a sectoral strategy for water services.[397] In the domestic sector, the strategy differentiates between the former white areas, and others such as the former townships or the peri-urban settlements.

---

[381] Water Supply Act s.319, and generally Part 9A.

[382] Water Supply Act ss.50–61.

[383] Plumbing and Drainage and Other Legislation Amendment (Qld) Act 2005 No.39.

[384] Department of Housing and Public Works (2013a) Queensland Development Code Part MP4.2.

[385] Department of Housing and Public Works (2013).

[386] Water Efficiency Labelling and Standards (Qld) Act 2005 No.69; Water Efficiency Labelling and Standards (Cwlth) Act 2005 No.4.

[387] Department of Communities and Local Government (2010).

[388] HM Government Building Regulations (2010) Parts G and H.

[389] Water Supply (Water Fittings) Regulations SI 1999/1148 Sch.2 para.25.

[390] European Commission (2012e) Chapter 6.

[391] Building Standards for Scotland Technical Handbook (2013) Domestic Part 3, Non-Domestic Part 3.

[392] Water Bylaws (Scotland) (2004).

[393] New offences have been created recently, for commercial users; Water Resources (Scotland) Act s.35, to complement the existing general offence, Sewerage (Scotland) Act s.46.

[394] Scottish Water (2013).   [395] DWAF (2005).

[396] Department of Public Works (2000).

[397] DWAF (2004c, 2004d). There are also sectoral strategies for agriculture, industry and mining.

The paper recognises that, since reform, there has been less emphasis on providing for the 'former white areas' and therefore leakage and unaccounted-for water may have increased in these areas. There is also recognition that these households, which use water at developed world levels comparable to Australian cities, contribute a high proportion of revenue, enabling cross-subsidy; WSPs may be reluctant to reduce consumption by these users. In the townships, there continues to be very poor quality infrastructure and fittings and significant wastage, whilst in the settlements and in rural areas the challenge is to provide a supply at all. The documents do provide a framework within which conservation and demand management can be mainstreamed into thinking and practice amongst both institutions and users, and demand management is a focus of the NWRS2,[398] but without specific legislative measures they will remain aspirational policy statements. There is little appetite for retrofit in any of the jurisdictions, and that seems least practicable in South Africa.

5.6.4 Stormwater and sustainable urban drainage systems

Runoff from roofs, streets, car parks and other impervious areas is often disposed of to a separate network of surface water drains, discharging directly to controlled waters. Traditionally, stormwater, and other surface runoff, is managed by roads and transport authorities.[399] It may carry potentially polluting substances, such as hydrocarbons from vehicles. It may also be discharged to sewers, either through a single system or due to misconnections, in which case there is likely to be sewer flooding in the event of heavy rain. Depending on the climate, it may be desirable to restrict or discourage the use of impermeable surfaces, to prevent runoff, and/or to restrict water-intensive plants. A number of engineering techniques allow the filtration of runoff into the soil through permeable surfaces; some of the approaches to building control discussed above, for management of greywater, are also relevant to surface runoff.

In the UK, Scotland led on introducing Sustainable Urban Drainage Systems (SUDS), and these were given a legal basis, with a statutory definition and provision for their subsequent adoption and maintenance by SW.[400] Scotland has also been developing a partnership model for major urban drainage schemes,[401] involving SW and all the local authorities as well as SEPA and the Scottish Government. A new Floods Act has imposed duties of cooperation and established powers to address urban drainage.[402] In England, there has been much more resistance to SUDS, with concerns over ownership and maintenance; but following major flooding in 2007 and 2008, enabling powers were enacted.[403] There has been a consultation on draft standards, but so far no orders have been made.[404] In Queensland, there is extensive guidance on stormwater drainage, including SUDS-type systems.[405] Until recently, local governments were required to establish a Total Water Cycle Management Plan under the Environmental Protection (Water) Policy.[406] This has also been repealed as part of the deregulation agenda,[407] but guidance remains and the policy concept is referenced widely.[408] In South Africa, guidance on building standards for water and wastewater includes guidance on stormwater management.[409] As raw water becomes more valuable, the calculations around the economic level at which treatment and reuse is viable will change; there is surely an argument for looking again at institutional responsibilities here, as this separates out management of part of the water resource, with implications for quality and quantity.

5.7 CONCLUSIONS

This chapter has examined the law regulating the delivery of water and sanitation services. Ownership, industry structure and regulation all have an important role. Ownership may be in the public or private sector, and this will affect the source of capital. Regulation may be through a government department or an independent regulator, and this choice does not depend on the ownership model. Whether or not there is a separate regulator, a competition authority will be well placed to exercise some supervision and review. Further, the same regulatory model can be applied to both private entities and public organisations, although the focus may be different. A public organisation may tend to inefficiency, a private entity to excess profit.

Water services law is in many ways the most difficult to reform. There is unmet need for both rural and urban poor, but the areas in greatest need are least attractive to private sector investors, and the regulatory environment is complex. The problem is compounded by the strength of political views, and the difficulty with taking a nuanced stance on any aspect. The political dimension is crucial. In many countries there is strong

---

[398] NWRS2 Section 3.3.3 and Chapter 7.

[399] Local Government (Qld) Act 2009 ss.66–80; Constitution of South Africa Sch.5 Part B; Highways Act (UK) 1980 c.66 s.100; Roads (Scotland) Act 1984 c.54 s.31.

[400] WEWS s.33; although there continue to be difficulties with the adoption of systems constructed before formal design criteria were developed.

[401] 'Metropolitan Glasgow Strategic Drainage Partnership' see http://www.mgsdp.org/.

[402] Flood Risk Management (Scotland) Act 2009 s.1 (cooperation); s.17 (SUDS).

[403] Floods and Water Management Act 2010 s.32 and Sch.3.

[404] DEFRA (2011a).   [405] DEWS (2013a).

[406] Water Policy 2009 ss.19–20, to include management of sewage and stormwater.

[407] Water Amendment Policy (No.2) 2013 SL No.272.

[408] DERM (2010b).   [409] Department of Public Works (2000).

opposition to private sector involvement; a presumption in favour of the public sector may alleviate concerns. For many countries, significant private sector investment will not be the answer.

In every country government sets overall policy, including social and environmental goals as well as service standards. Government must provide adequate regulation – that is its core function and cannot be delegated. The regulation of prices and service standards may be separate from control of drinking water quality, and environmental issues (abstraction, discharge of wastewater, management of solid wastes), but the price setting process must be wide enough to identify the costs of these activities and ensure that they can be achieved for the price set. Governments may wish to set prices if supply is in the public sector, but it is possible to have independent regulation of prices even with a public supplier.

Given the natural monopoly characteristics, it might be assumed that a vertically disaggregated sector, at least with separate bulk supply, would be inherently more competitive. Yet, in South Africa and Queensland, there is still little direct competition at the level of retail and distribution, and no particular policy agenda to introduce the same. In Scotland, there is some minimal competition that has been effective; in England, wider competition is being promoted, but subject to criticism and concern. In a classic economic analysis of natural monopoly, regulation is a substitute for competition; but perhaps it is better to concentrate on effective regulation, where much can be achieved. In England, the regulator has struggled to maintain effective control over the financial structures of firms. In Scotland, there have been significant efficiency gains, but perhaps equally important has been the emergence of a mature public governance ethos, where the provider, the government and the various regulators are genuinely cooperating. The Scottish model will not work everywhere, but some lessons from Scotland appear valid, including the benefit of explicit regulation within the public sector. The legislative framework is much less voluminous and relatively clear compared to England.

Disaggregation will necessitate contractual arrangements, regardless of whether there is any PSP. If contracts are used they should be publicly available and Ministers should have a power to impose mandatory terms, certainly for bulk supply. If any participants are in the private sector there may need to be specific provision for publicising the contracts, as freedom of information laws may not apply. Contracts should always be subject to regulatory oversight.

There are arguments for a vertically aggregated sector. This may leave regional monopolies, but benchmarking is one regulatory option that can be successful. Vertical integration avoids the need for contracts, although licences may be required instead; again these should contain standard terms and be published. This structure may be better placed to pay proper attention both to upstream catchment management and to downstream wastewater management, both of which are becoming increasingly important.

The human right to water is the subject of much attention. In South Africa, the existence of a constitutional right gives avenues of enforcement and protection not available elsewhere. Whilst a constitutional right may open up additional legal avenues, it is not a substitute for a properly specified provision in a sectoral rule. In developed countries, almost everyone will have an adequate supply, though price rises may cause some hardship given other pressures on household income. Some countries allow disconnection or reductions in flow for non-payment or other breach, which is a political decision; some provision for health and hygiene must still be made.

Social tariffs are an important component of the pricing regime. The full cost recovery principle is a starting point, but a presumption of unwinding all cross-subsidy is not practised. It is likely that there will be cross-subsidy for social equity, from industry or high volume domestic users, or between urban and rural; the policy framework should specify this. There may also be subsidy from general taxation, which is best focused on bulk infrastructure and on network extension and connection. The problems in the developed and developing world are different; it would behove us well, in the north, to remember that our water and wastewater systems were, for the most part, initially provided by the state. Block tariffs are likely where there is metering; arguably, in some South African municipalities the volume of the FBW allowance is set rather low, relative to the pricing of the second block.

It is not necessary to have punitive cost recovery as part of a more commercial approach, and a more commercial approach does not need to be part of a 'privatising' agenda. Traditionally the public sector may tend to inefficiency, and traditionally water services are provided by local governments with multiple functions. A more commercial approach, financially ring-fenced, enables better asset management and future planning and assists with transparency; in turn, it should make it possible to provide a better service and to make better provision for the poor and unserved. It is the conflating of a more commercial approach to costs and liabilities with limitations on social tariffs and harsh debt recovery that obscures the arguments here.

The proper representation of the customer interest is a vital part of good governance. Clarity of the legislative framework and transparency of data are vital; in most of the laws examined here the legislative and policy contexts are extensive and complex and the raw data difficult to understand, even where they are available. General and specific duties will apply to service providers and may be enforceable only through the regulator, or by individuals as well; the latter gives more options for accountability, but may still be constrained by the broader regulatory environment. Often customers are represented by the regulator;

separate bodies may be established when the regulatory framework matures, but mechanisms to engage with customers and seek their views on trade-offs between service standards and price are always necessary.

Regulatory institutions do not depend on ownership of the assets. In Scotland, an independent regulator sets prices for a public entity; in South Africa and Queensland, government controls the price of bulk water, and monitors, but does not set, retail prices. A multi-utility regulator or competition authority may have a regulatory or supervisory role. Regulation may also be cyclical. The evidence in England and especially Scotland is that a 'lighter touch' may be feasible, but only after a period of quite intense regulatory effort in which significant datasets are identified and examined. In Scotland, there is a genuine success story in regulation and governance of the public sector, which will continue to be the location of the delivery of most water services.

Queensland has seen two sets of extended reforms in a decade. The first responded to drought by creating a series of institutions and responsibilities. The second removed many of these, following a deregulatory agenda and after the drought broke. It is unlikely that many other countries would wish to move that quickly. Queensland decided, in some ways similar to Scotland in the 1990s, to remove water services from local government control and establish other corporate forms – but only in the urbanised SEQ. For a jurisdiction with such a small population to run two parallel systems of service provision, with different regulation and governance, has meant some operational difficulties, but the Water Supply Act is well structured. It might have been clearer to have all regulatory matters in that Act, and retain the SEQ Water Act for the governance arrangements for the retailers and their constituent councils. In South Africa, the WSA is clear enough, but the accompanying regulations are rather minimal and need to be revised. The surrounding policy is as complex as any of the other jurisdictions.

Ultimately, if the resources are not available then some levels of service cannot be provided to all, and here the engagement and participation processes analysed in Chapter 2 are again important. From the design and maintenance of rural sanitation, to the trade-offs between cost and levels of treatment in small supply zones, stakeholder participation is vital; both the lessons learned, and the practical mechanisms used, in establishing catchment planning may also assist in water services provision.

The focus must be on effective regulation of prices and standards regardless of the ownership of the provider. In terms of the technical standards for service provision, a modern water law will not depend on qualitative descriptors such as 'wholesome'. There will, and should, be technical parameters for acceptable drinking water quality. Where resources are scarce, thought should be given to a reduced set of parameters more easily tested

and enforced by smaller authorities. This may be achieved by the use of guidelines rather than mandatory standards, or by allowing some deviation from standards, but in both cases with appropriate guidance. As regards sanitation, in rural areas currently unserved, serious consideration should be given to supporting alternative methods of disposal of human wastes, but these seem much less appropriate in urban areas. Amongst the complexities of business planning, asset management, financial reporting and technical auditing, the duties to provide the service are paramount and the mechanisms for securing their performance should be just as clear as those for economic regulation.

In all the jurisdictions considered here, there are high level duties regarding efficiency and water conservation at the level of providers and regulators, and a prohibition on waste, or alternatively a duty to use efficiently, on users of the service. These will rarely be invoked against individuals, but set an appropriate context for policy on conservation and demand management. They may be established in an overarching law on water resources, but are likely to be separately specified in water services legislation. Use of greywater systems, and other reuse of wastewater, is one way to conserve resources, but is only effective with full support from all the regulatory agencies. Again, the framework and the principles may be found in wider water resource management structures. If water efficient practices are mandatory in domestic housing and in industrial and commercial operations, if the price of raw water rises to a level that makes the sale of treated wastewater viable, then change will happen. This does require joined-up legislation; if the environmental regulator, or the planning authority, does not support change it will be less effective.

This chapter has shown the links between water services law and the other regimes examined in this book, and also the differences, particularly the role of economic regulation. Bulk water supply, or the water services functions of irrigation authorities in rural areas, could be incorporated into legislative reform of water resources management, as happens to an extent in Queensland and in South Africa. However, if the intention is to reform the business models and service standards for urban water services, that is a very different law reform package that should be conducted separately from the reform of IWRM and, within that, of abstraction and pollution control.

There is no right answer to the best way to organise water services, and this book has not sought to develop a single best model answer in any of the areas it has examined. The important issues are clear governance frameworks, clarity of the prevailing policy context, and direction provided by government as to provision for the poor and meeting environmental as well as service goals, with accountability for service providers. The protection of the consumer should be at the heart of a modern water services law; for life, and for human dignity.

Table 5.1 *Key findings Chapter 5: water services*

	South Africa	Queensland	England	Scotland
Regulators:				
Economic/ duties of supply	Department. DWA.	Department, DEWS; Competition Authority.	Agency. OFWAT; Competition and Markets Authority.	Agency. WICS; Competition and Markets Authority.
Environmental	DWA.	DEHP.	EA.	SEPA.
Vertical integration	No. Water boards; local government.	No. Bulk suppliers; local government; other public bodies.	Yes. Regional monopolies.	Yes. National monopoly.
Ownership and PSP	Almost all public (local government). PSP permitted. Limited concessions/contracts.	Almost all public (local government, rural water authorities, urban joint distributor-retailers). PSP permitted.	Private monopolies; licensed. Some direct competition, being expanded.	Public corporation; BOO schemes for wastewater. Limited retail competition.
Constitutional right	Yes, progressive.	No.	No.	No.
Duties of supply	Universal obligation (basic services); progressive access: efficient, affordable, economic, sustainable.	Duty to supply in service areas; greatest practicable extent.	Duty to supply (domestic); subject to high level duties (SS, OFWAT). Consumer objective; competition; secure functions; return on capital; economy and efficiency; sustainable development.	Duty to supply (domestic) at reasonable cost; sustainable development.
Metering and disconnection	Presumption; not from FBW.	Presumption; restrict supply.	Water-stress areas; no restriction/ disconnection.	Not domestic; no restriction/ disconnection.
Emergency powers	General powers; reasonable steps.	Drought, contamination, infrastructure failure.	Drought Orders and Permits; emergency supplies.	Water Shortage Orders; emergency supplies.
Conservation/ water efficiency	Effective management and conservation.	Sustainable management and efficient use; safe and reliable supply.	Conservation; efficient use.	Conservation and effective use; adequate supplies.

6   General conclusions

6.1   BEGINNING THE REFORM PROCESS

This book set out to explore whether it was possible to develop a framework for reform of national water laws, drawing on emerging practice in four jurisdictions where the law has recently been reformed. The conclusions are set out below; findings from each chapter were expressed in table form at the end of each chapter, and the overall findings are also expressed in this way (Table 6.1). The purpose of these tables is to provide a framework against which others can analyse subsequent reform processes.

The conceptualisation of the meta-regime of water law (Chapter 1, Fig. 1.1) was the first step. Whilst this book has for the most part considered only the principal operational regimes, there will be links with the other aspects of water management, as well as other strategic meta-regimes.

This conceptual process will also assist in determining how the principal operational elements should be grouped together. For example, if water pollution is already, and is to remain, a part of a different environmental regime, then the option of fully integrated water use licences will not be available.

A series of preliminary questions, developed from the research undertaken, will elicit guidance in the preliminary design stages:

* Is there a system for water resource management in place?
* Is it the intention to introduce or further reform such a system?
* Which department or agency is to be the principal regulator – for resource management, and for each of the principal operational areas?
* Are the functional areas of abstraction control and pollution control also to be reformed?
* Are there to be integrated water use licences, and if so, which uses are to be combined?
* Is there to be reform of water services law?
* If so, is that only for reform of bulk supply and/or rural water services, or is there to be reform of the whole structure and delivery of urban water services?

6.2   WATER RESOURCE MANAGEMENT

Although the sequence of chapters in this book is postulated as being the most logical progression, it may be that not all elements currently require reform, or alternatively, that there are pressing reasons for beginning a reform process with one of the functional areas. However, if there is no legal framework for resource management, or if the same is deficient, then it is likely to be most effective to begin at that point. In any event, the conclusions follow the order of the book, without any stipulation that such is an essential progress.

As regards the IWRM process, certain features are accepted good practice and are likely to be adopted in any jurisdiction; particularly, using basins or catchments to set management boundaries, and the need to integrate the management of groundwater with surface water. A series of decisions then need to be made as to how the new planning system will be operationalised. Preliminary questions here are whether the process will be led by the regulator or by the stakeholders; whether a special board or agency will be constituted; and whether licensing functions will devolve to such a body. The choice between departmental and other regulators is a wider political question.

It is necessary to consider the structure of the proposed legislation itself. It is likely that there will be separate legislation for water resources and water services; but bulk supply, or rural water services, through, e.g., irrigation networks, may be within a water resources law. Environmental law is a whole separate meta-regime and its structure may affect the options for water management. The legislation in South Africa is clear and easy to use, though the secondary legislation in water services is in need of some development. During the reforms in Scotland, water resources law was rationalised; that is less true in water services, but the legislation is generally clear. In England, the two principal Acts are clearly structured, but the actual legislation is much-amended and difficult to read and use. In Queensland, the extensive institutional reforms have in some cases left two sets of applicable rules. Reform can provide the opportunity to rationalise prior law and clean up the statute books, and such is welcome. Legislation that is clearly expressed and easily comprehensible is a desirable goal for legislators and drafters alike. Indeed, a useful design consideration might well be to engage as few legal instruments as possible in any sphere of reform.

Policymakers can then consider the necessary content of the legislation; the following questions should be considered for the development of the primary law. Again, these are design questions that have emerged from the research undertaken:

- Will there be any express provision for ownership, trusteeship or vesting of the resource in the state?
- What will be the purposes, or principles, or high level duties, to be stated at the outset of the legislation?
- What will be included in the definition of the water environment?
- What will be the designation and role of any appropriate authorities, and the delimitation of boundaries for planning purposes?
- What process will be prescribed for the production of the plan or strategy?
- What provision will be made for stakeholder engagement?
- Will there be explicit provision for the status of the plan in relation to other statutory functions and/or other planning processes, and any duties of cooperation?

Although this book did not set out to make recommendations as to the best model or option, as any reform must be appropriate to the prevailing social, political, economic and legal contexts, some findings seemed sufficiently clearly established to merit a recommendation. One such recommendation is that a state introducing IWRM uses primary legislation, in an act or code, as in every jurisdiction here save England. This enables contentious decisions as to principle to be properly debated, and in turn increases legitimacy and minimises the risk of significant opposition. Two matters best addressed in this way are any provision for state ownership or trusteeship of the resource, and any high level principles, purposes or duties established regarding water resource management and use.

Explicit provision need not be made for ownership of the resource, and its omission may avoid contentious discussion; an alternative view is that it gives clarity in establishing the state's role and responsibility. Once established, then conflicting concepts of prior property rights are less likely to strike at the fundamentals of the new regime. If enacted in the strategic law of IWRM, concepts of state trusteeship or ownership, and high level principles or duties, can then apply to each of the subsidiary operational areas.

Underpinning principles for water resource management include equity, water efficiency and/or conservation, and sustainable and/or beneficial use. These depend in part on the prevailing socio-political context. For example, where there are significant inequalities in access to resources, equity is likely to

be a priority. In Queensland, there is an overarching statutory purpose, further defined in a set of principles, which can then apply to decision-makers, users or service providers. There are similar principles in other jurisdictions, but in Queensland there is a high degree of specification, providing criteria by which to assess the performance of functions.

As regards definitions, it is assumed that a reformed water law will incorporate both surface waters and groundwater, and the legislation should include both in the definition of the water environment, in order to protect the whole resource adequately. The incorporation of (near) coastal waters or wetlands is desirable, if by no means essential.

There are also questions of scale. Managing resources at a regional scale may enable concentration of resources, but will need input from local sub-catchment organisations; this should be foreshadowed in the stakeholder engagement provisions. Only South Africa attempts to devolve regulatory authority to stakeholders.

Wherever possible, planning for other areas of water management (coastal waters, dams, and floods and drought) should be linked into IWRM plans, perhaps as sub-plans. Plans or strategies for some sectoral uses of water may also be integrated in this way, depending on their relevance in the area, particularly mining, agriculture, hydro or navigation. As regards other existing planning processes, such as land use, governments are unlikely to make express provision as to their relative status. Nonetheless, there should be duties on all relevant authorities to at least consider the water management plan. There may be a duty to further the objectives of the plan or strategy in carrying out functions under the principal Act, as in South Africa, or in carrying out other functions, as in the UK; these may be imposed on all relevant authorities or specified agencies. In Queensland, there is a duty to exercise powers and functions to advance the overall purpose of the Act. Queensland also makes the most comprehensive provision for liaison between authorities at the level of project decisions.

6.3 WATER RIGHTS AND ALLOCATION

The reform of abstraction rights may be controversial and attract resistance from parties with existing rights. In a staged process, IWRM will provide both mechanisms and experience for stakeholder engagement, but if there is no wider reform, or if the regional scale is too large, then efforts must be made to engage with key sectoral users and local catchment groups, such as WUAs or local irrigation boards. This will be particularly important if there are high levels of agricultural abstractions by small-scale users. A reformed law will almost certainly utilise a licensing regime for abstractions, and it is possible that the law of dams will be partly or fully integrated with these rules.

If there has been legislative provision as to public ownership or trusteeship of the water resource, or high level duties or principles, these will provide a helpful context, underpinning any reallocation. If water rights are being reformed alone, then this will be an appropriate opportunity to consider making provision as to ownership or trusteeship, and certainly to establish general duties on both authorities and users. If there are constitutional rights to property, then care must be taken to ensure that these do not inhibit the reallocation of water or lead to challenge of the new regime or regulatory decisions within it. It is difficult to make general recommendations, but acting proportionately, in accordance with properly adopted primary law, with public notice and a long lead time, should avoid allegations of expropriation. Again, early stakeholder engagement is critical to ensure legitimacy.

It is unlikely to be appropriate to retain only user-controlled access to water, such as riparianism or prior appropriation; but customary systems should be given special attention. Consideration should be given as to how to deal with existing users, with special provision for small abstractions, on grounds of equity, to meet basic human needs and to avoid an unmanageable regulatory burden.

There are various possibilities. The smallest scale, often domestic or subsistence uses, may simply be exempt, as in South Africa, or subject only to general rules, as in Scotland. The lesson from Scotland is that even in a small and well-resourced jurisdiction it is too expensive to even register, let alone license, every water use. England and Queensland also effectively authorise small abstractions, either below a volumetric limit or for stock and domestic use, but it is still important to provide some mechanism to bring these within control if need be. It is not desirable to grant extensive protections for existing users making larger abstractions. This can create problems later, when pressure on the resource increases, as has happened in England, and, given current uncertainties over climate change, seems a high risk approach. It would be better to take time to engage with existing users and achieve a broad consensus for a fuller licensing regime.

Scotland and South Africa also make extensive use of a simplified consent procedure for all uses of water, not just abstractions, within specified limits. These uses are registered, and subject to general conditions in the regulations, and may last indefinitely or for a statutory period. Such general authorisations are a very useful tool that can redirect regulatory effort to monitoring and enforcement.

For the largest abstractions, and other large-scale uses if integrated, a full licensing regime will be appropriate, with applications advertised for comment. The regulator should have powers to serve enforcement notices in whatever form the jurisdiction employs, the usual powers of access, inspection and investigation, and powers to review and if necessary revoke licences. For clarity and certainty, and to avoid challenge, criteria for granting licences, and grounds for review and revocation, should be specified in the legislation. It is not recommended that licences be perpetual, to allow for adaptive management, and there seems no good reason not to specify time limits for licences, and time periods for review, either in primary or secondary law. The legislation should set out a clear procedure for review and appeal, making use of any specialist bodies as appropriate. If appeals are made to Ministers or subject to internal departmental review, there should also be opportunity for further appeal to the courts as appropriate to the legal system in question. The South African experience shows that it may be necessary, for equitable reasons, to compulsorily reallocate water licensed under a previous regime. An objective of reform of water rights should be a well-designed permitting system, with appropriate constraints, that does not allow the establishment of permanent rights in water and hence the need for such contentious procedures in the future.

The decision whether to firstly permit, or secondly encourage, water trading is a political decision. Globally, few states have adopted the position whereby allocation is left solely to the market; the approach taken by all the comparators here is to allow trading within a planned system. Beyond that, it is feasible to limit trading to irrigation, e.g., within irrigation districts, or to facilitate it more widely, across districts, or basins, and across sectors. Scotland and South Africa have legislative provision to enable trade whilst retaining regulatory control, and England's ongoing reforms are specifically to make trading simpler and less costly. The proposals in South Africa to move away from trading, and also to institute a 'use it or lose it' principle, reflect the pressure on the resource and the ongoing need for equity. In Queensland, the policy context is more advanced, as is the incidence of trading, and the legislation reflects this. It provides for separation of water rights from land, with appropriate protections for existing security holders, and especially for a secure system of registration of water rights. There is protection for owners of distribution networks against stranding of assets. The rights themselves are allocated periodically under the statutory water planning system, and there is a new formula for sharing risks in future in the event of reduced availability. All of these would be useful models to any state seeking to advance the use of markets, but the priority for reform is likely to be, firstly, a sound system of resource management and within that, of allocation; and, secondly, the decision to permit trade and facilitate it. Beyond that, there may or may not be a socio-economic environment such as to mandate more permanent and extensive use of economic instruments.

6.4 WATER POLLUTION AND WATER QUALITY

The preliminary design questions will already have indicated whether water discharge consents will be integrated with other uses, or whether they belong in a different meta-regime. If the latter, then water regulators should be consultees on discharge consents; if the former, then there will likewise need to be a system of referrals with any environmental regulator, especially for licensing waste to landfill.

The conclusions in Section 6.3 above on licensing regimes for abstractions will also apply, for the most part, to discharge consents, in terms of criteria for grant, review and revocation. If there are high level duties or principles applying to all uses of water, these will apply to discharges. There may also be general environmental obligations on all citizens and, perhaps, statutory provision for citizen review of regulatory decisions and other matters affecting access to environmental justice. Pollution offences are likely to be strict liability offences and, again, there will be standard investigatory powers. A wide range of administrative penalties within the control of the regulator are likely to be more useful than relying solely on criminal convictions.

The law will inevitably use some combination of emission and quality standards. The former will apply to discharge consents, and the latter will be used to calculate discharge consents and also as a check on other diffuse sources of pollution. In reforming pollution control, there may be fewer jurisprudential difficulties, as the prior law is unlikely to have manifested 'rights to pollute' in the same way as rights to abstract. Hence this area may be more susceptible to incremental reform. Basic effluent standards are a starting point, variable in more sensitive waters. It is also likely that there will be a set of environmental quality standards, and these will apply to waters depending on the use to which they will be put – drinking, irrigation and bathing or recreation are obvious use classifications. Where regulatory resources are scarce, it is better to have a smaller set of core parameters than an extensive range that cannot be monitored or enforced. The UK experience following EU membership in the 1970s suggested that mandatory standards are more effective than guidelines or other regulatory discretion, but guidelines are an alternative. A narrow set of standards and a wider set of guidelines are one possibility. In Queensland, guidelines may be departed from if water quality is still improved.

All the jurisdictions are increasingly focused on ecology, and protection of species and habitats, measured by the range and health of aquatic ecosystems, with a consequent ecological classification system. This involves managing not just flow, but also the physical structure of the river and riparian zone, as well as achieving chemical and physico-chemical quality standards. This wider focus shows very clearly the importance to water quality of effective resource management, as these requirements cannot be implemented through individual environmental permits alone. They require the joined-up processes, and the data, found in IWRM. Similarly, if groundwater is already part of the applicable definition of the water environment, then the ingress of contaminants can be controlled through environmental licensing, but broader questions of groundwater quality, and control of diffuse pollution, must be managed holistically. Catchment management should allow integration of control regimes, and also enable and facilitate better interdisciplinary work, at the interface of law, science and policy.

There are differences between the jurisdictions in their ecological protection systems, but also many similarities. The EC WFD sets a mandatory target of 'good' status, but with extensions or exemptions based around cost, feasibility and human needs; the process is narrated in the RBMPs. It would be fair to say that the WFD process is a massive restructuring and also rationalising of 35 years of European water law.

Both South Africa and Queensland have benchmarks for 'ecological water' built into the quality guidelines, and a similar classification system to the WFD, but neither prescribes a particular overall goal for all waters. This may be a more honest and less complex approach than setting an overall goal and then allowing exemptions.

Significant reforms in Scotland and South Africa were enabled by prioritising the water agenda, through *inter alia* integrated water use licences. This detaches water reforms from dependency on other reform agendas and timetables, and therefore, subject to context, this is also a recommendation.

6.5 WATER SERVICES

Whereas water allocation and pollution control can be entirely subsidiary to an IWRM framework, water services law cannot be planned for nor implemented through IWRM alone. Major upstream activities, such as large dams, significant abstraction points, and water transfers, will certainly be a focus for river basin planning. Water quality issues, especially wastewater management, but also catchment protection, will also be part of IWRM. Further, WSPs of various sorts will hold licences to impound, to abstract, and to discharge along with other water users, and may be bound by the same duties and standards. But water services will inevitably require separate strategic and operational planning of a different sort, and are also subject to extensive service-specific obligations, both social and technical, which may be created in statute or contract or both, although all the jurisdictions here take a mainly statutory approach.

If it is the intention to make widespread reform of the structure of the water services industry, this is a separate reform agenda which is not recommended to be undertaken concurrently with reform of the water resource management agenda. It is possible to incorporate the supply of bulk water into resource management legislation, and perhaps the establishment of rural water institutions that also provide water services, but urban water services involve quite different areas of commercial and corporate law and should be considered separately.

If there are overarching principles, purposes or duties in a national water law, these can apply to service providers as well as other actors, including where services are provided by local government. Inevitably, local government providers will have separate regulatory and reporting systems, but their reporting on water services should find its way to the water services regulator. If there is a constitutional or human right to water, it will provide additional mechanisms for enforcement, but the human right to water has broad acceptance; states must provide these services regardless of constitutional provision. The content and extent of the duty to supply must be expressed in sectoral rules, and these will in turn assist in defining and implementing the constitutional right.

The difficult questions arise at the start and continue. Water services are capital intensive with long-term investment needs. There is a reluctance to acknowledge, and pay for, the true cost of provision. There are many developing countries with extensive unserved populations that cannot obtain the capital funding required to serve them, at least without significant price increases. Furthermore, it is a service where proponents of both sides of the public–private dichotomy hold their opinions with a religious zeal. Perhaps the first critical question should be: What structure is proposed for ownership? But in practice that first question would be more accurately expressed as: What structure is proposed for ownership that will be politically acceptable?

The question of ownership cannot be addressed without assessing the current structure of the industry and, particularly, whether there is vertical integration. A disaggregated structure may be *de facto* more competitive, as there are more opportunities for diversification. If private sector participation is a policy goal, then disaggregation will make it easier to introduce the same in some functions, whilst maintaining at least the networks in public control. However, vertical integration may give a better opportunity to manage from the catchment to the sea, and comparative competition has been reasonably effective in the UK. If water services are delivered by local government, then it is almost certain that some other structure will be used for bulk supply. There may be good reasons for removing water services from local government, but again that is a political choice with constitutional implications in many states.

There are many approaches available and utilised around the globe, some public, some private, and some a mix of both: in the public sector, integrated provision by local authorities, or by regional water boards; in the private sector, long-term leases and concessions; joint ventures and partnerships; and all sorts of smaller contracts for individual projects and elements of services. It would be helpful, if possible, to adopt a non-dogmatic position when reviewing the possibilities; if services are currently undercharged then prices may have to rise whether in the public or private sector. The use of the local private sector may be less controversial. Perspective is gained by recalling that the global fraction of water services provided for commercial gain is less than 15%. The basis of water services is public provision, and the focus of the global community must be the proper regulation of that public sector.

There is no one and best model. The jurisdictions surveyed here have a national public monopoly; a set of regional monopolies entirely in private ownership but licensed and heavily regulated; and two disaggregated systems, with various mainly public bodies, exercising bulk and strategic abstraction and distribution roles, and service delivery through local government and various other public, private and community bodies. Scotland, South Africa and Queensland have all made use of build–own–operate type schemes. Whether the industry is public or private or mixed does not alter the need for economic regulation. In Scotland, there have been excellent results applying such regulation to a public body; indeed, it is suggested that this is working better than in England.

Regulation may be carried out by a department, by a separate sectoral or multi-utility regulator, overseen by a competition authority, or use a mixture of these; but it must be wide enough to enable the achievement of environmental and social objectives, within a clear policy context that will be set by government. It is suggested, despite the classic economic view of regulation as a substitute for competition, that a focus on regulation as such can be very effective. It is also suggested that a more commercial approach to delivery in the public sector can bring great benefits, in terms of clarity around costs and asset management. This should be seen as separate from cost recovery as such, and especially subsidies, social tariffs and policy on debt recovery. Rather, it is a way of establishing the data necessary to operate the system, and to enable informed decisions on provision for the poor and unserved.

Many other questions in water services law are independent of the politics of ownership, and here there might be an argument for starting at the bottom of the legislative pyramid, with questions as to the appropriate set of quality standards for drinking water, and basic standards and best-practice approaches for sanitation and wastewater management. These will be less politically contentious, and might build consensus and improve stakeholder relations, although any significant improvement of standards will certainly come at a cost, and still raise difficult questions of the appropriate levels and acceptable trade-offs.

In the UK and Queensland, there is a presumption of full cost recovery in water services – although in both Queensland and Scotland, there is a substantial state sector, and some subsidy in rural infrastructure and deep network extensions respectively. Even in the UK, not all cross-subsidy will be unwound. In South Africa, there is a Free Basic Water service and a developing basic sanitation provision, although cost recovery is operated for other users and at bulk scale. Subsidy from central government, as well as cross-subsidy, both from business and from high tariff domestic users, is an accepted part of revenue. Queensland and South Africa have a presumption in favour of metering and both can reduce supply for non-payment. In every jurisdiction there are powers to restrict service in drought or other emergency, but that is very different from permitting restrictions for non-payment, which again is a political decision.

If local government is the existing provider, and the reform agenda does not envisage significant change, then larger authorities at least could adopt some financial separation to avoid this. Such commercialisation does not need to entail a harsher approach to debt recovery, but should instead enable transparency as to the source and destination of cross-subsidy. Alternatively, general tax and benefits systems can be utilised to protect the poor.

The other focus of legislation, but also of policy and education, should be instilling an ethos of conservation of resources and efficiency of operation. These twin principles, applied by providers and users alike, must underpin any possible solutions to water services for the uncertainty of the twenty-first century. These principles are, and should be, relevant to the wider arena of resource planning, and applicable to water abstraction and pollution control, as well as water services. They establish the sustainability dimension, reflecting the tripartite demands of environmental care, social wellbeing and engagement, and economic management. In the water services arena, water recycling and reuse, efficient buildings, innovative sanitation, sustainable drainage, and an aware citizenry all have a part to play; the first four can be legislated for. Small steps as well as large projects are needed, and a heavy dose of political will.

6.6 AN ANALYTICAL FRAMEWORK FOR REFORM OF NATIONAL WATER LAW

This book did not set out to develop theory. A feature of modern water law, and other developing areas of modern administrative law, is that they do not necessarily engage in theoretical legal debate. There are ideological arguments over water – for example between public and private service provision, between a rights-based and a regulatory approach, or indeed the merits of law against custom – although these differences may only divert effort from policy goals. There are also theoretical analyses. Economics, for example, provides a theoretical perspective, but often the counterarguments are not economic but political, social and environmental. The debate is not engaged on the same terms. Similarly, a property law analysis is a theoretical basis for discussing water rights, but the licensing regime that replaces private rights does not always answer that theoretical construct. It is simply imposed, with a pragmatic public law justification. That is not to say that, in such a process, efforts are not made to engage those stakeholders who hold a different perspective; indeed, that is essential to avoid continued resistance. But there is not necessarily an attempt to construct a different property-based or rights-based theory; again, the argument is not made on the same terms. Therefore, this book did not seek to make a theoretical argument. Rather it set out to make a positivist and pragmatist analysis of water law, in the normative context of its policy drivers, and thereby establish a framework for analysis of prospective water law reforms. That pragmatic approach underpins modern regulatory systems.

It is also hoped that the book has demonstrated the importance of the role of law in water management. As noted in the introduction, too often the law is perceived as essentially negative, and the role of lawyers is only to adjudicate – at great expense – once conflict has emerged and cannot be resolved by other means. Such a role will of course remain, but well-designed legislation should be able to provide a framework for action that is acceptable and clear to all parties, setting out rights and responsibilities, providing for information flows, facilitating stakeholder engagement, and with accessible mechanisms for review and enforcement where conflicts do emerge.

The intention was to examine whether it was possible to extract a framework of the core components of any national water law. Although the four comparators use different mechanisms and institutional structures, there is a high degree of commonality as to what is being done, and, to an extent, how it is being done. The final conclusions contain more normative elements, in the shape of recommendations and suggestions for best practice, than was envisaged at the start. It is hoped that it will make a useful contribution to ongoing work in many countries.

Table 6.1 *Overall conclusions and recommendations*

Reform area	Conclusions and recommendations
Preliminary design questions with regard to the overall proposals for water law reforms	Is there a system for water resource management in place?
	Is it the intention to introduce or further reform such a system?
	Which department or agency is to be the principal regulator – for resource management, and for each of the operational areas?
	Are the functional areas of abstraction control and/or pollution control to be reformed?
	Are there to be integrated water use licences, and if so, which uses are to be combined?
	Is there also to be reform of water services law?
	If so, is that only for reform of bulk supply and/or rural water services, or is there to be reform of the whole structure and delivery of urban water services?
Preliminary design considerations applicable to the introduction or reform of IWRM (also relevant to reform of allocation and/or pollution control)	It is recommended that primary legislation be used.
	Consider making express provision for ownership, trusteeship or vesting of the resource in the state.
	Identify the purposes, or principles, or high level duties, to be stated in the legislation (e.g., equity, sustainable management, beneficial/efficient use).
	If expressed in sufficient detail, and supported by procedural mechanisms, these may be capable of use as a check on regulatory and operational activity.
	Use the legislative opportunity to create an enabling statutory framework for the principal operational regimes of allocation and pollution control, if they also need reform.
	Use the water resource management framework and process to provide a point of interaction with other water law regimes (especially flood and drought, and coastal/marine law); and the principal sectoral uses of water; and integration with the related strategic regimes.
	Identify opportunities for rationalisation and reduction of multiple planning instruments and processes.
	Identify any international obligations, which could be expressed in the IWRM legislation.
IWRM: operational considerations	The definition of the water environment:
	It is expected that there will be a broad definition of water environment, including surface waters and groundwater; this may also include coastal and estuarine waters; wetlands; diffuse surface waters. This gives the greatest protection to the water environment for water resource planning and under the related functional regimes for abstractions and pollution control.
	The designation and role of the relevant authority, and boundaries for planning purposes:
	It is expected that river basins/catchments will be used as the basis of administrative units. It may be helpful to state the status of IWRM plans or strategies relative to other planning processes (especially land use, conservation). It may be helpful to create duties of cooperation, and/or duties to consider IWRM plans, for other authorities (especially land use planning).
	The process for the production of any plan or strategy, and the responsibility for setting, achieving and reporting on targets:
	The process may be regulator-led or stakeholder-led; this may be linked to whether the plan has a regulatory effect or is purely managerial (e.g., in terms of making allocations, or granting licences; and/or whether any targets set under the plan are directly or indirectly binding).
	Provision for stakeholder engagement:
	Identify existing catchment management activities, formal and informal, which may be utilised.
	Identify key stakeholders/water users for that basin.
	Identify any existing participatory frameworks (e.g., in environmental assessment, land use planning) that may be utilised.
	Ensure that the detail of participation is set out in the applicable legislation, e.g., access to documents, specific timescales, mechanisms for the general public as well as water users.
	Specifically, require regulators to provide reasons for administrative decisions and access to background data. This should be applied to all aspects of water law.

Table 6.1 (*cont.*)

Reform area	Conclusions and recommendations
Water rights and allocation: preliminary design considerations	It is recommended that primary legislation be used, especially if there is to be reform of pre-existing water rights.
	If there is no wider programme of water reform, and specifically no IWRM legislation, consider making express provision for state ownership or trusteeship;
	establishing any high level purposes/principles (e.g., equity, sustainable management).
	Consider any specific duties applying to abstractors (e.g., beneficial or efficient use, or a no waste principle).
	Consider whether there should be a statement as to the priority uses of water; and if so, whether this should be applicable at all times, or only in times of shortage or other emergency; and whether such uses should be exclusive, or include other reasonable uses; and whether such priorities should be stated in the legislation or in policy.
	If a statement of priority uses is to be made, then the ecological use of water should be expressly included.
	Consider whether there are to be combined water use licences, for abstractions, impoundments, discharges and river works; if not, then abstractions and impoundments may still be regulated together.
	Consider whether there are specific uses of water that should be controlled relative to abstraction; for example non-native species, or certain forms of forestry.
	Consider the extent of provision necessary for bulk water abstraction and transfer (and links with the law of navigation and the law of canals).
	Consider in principle whether water trading is to be permitted, or encouraged.
	Recognise that the introduction of comprehensive controls on water use is a long-term project.
Water rights and allocation: operational considerations	Identify the appropriate regulator.
	Permanent licences are not recommended, as both limiting adaptive management and fostering the view of water rights as protected property rights, hence opening the regime to challenge.
	A tiered system of control is recommended.
	Consider what provision to make for small-scale and existing users, for equity, to manage regulatory effort and to avoid subsequent challenge to the regime.
	Such provision will include lead-in times; transitional provisions for existing licensed uses; any compensation provisions; and stakeholder engagement and dispute resolution mechanisms.
	Consider appropriate scope of exempt uses/limits for general rules, and/or any registration requirements.
	Ensure that there are adequate powers to bring either areas/water bodies, or exempt uses/ users, into control on specified statutory grounds (for example environmental harm or water scarcity).
	Establish appropriate licensing procedures backed by enforcement powers, with public consultation on at least major projects, clear statutory timescales and criteria for grant, review and revocation of licences, in the legislation.
Water rights: operational considerations for water trading	If water trading is to be permitted:
	It is recommended that water trading takes place within a planned framework for water resource management, and a licensing system for allocation.
	Consider the extent of permitted trading: within or across basins; within or across irrigation districts; for irrigation only; for all uses.
	If water trading is to be encouraged, consider:
	What provision is necessary for a secure register of rights similar to a land registry.
	How to provide for security holders in separating water rights from land rights.
	How to manage stranded assets.
	What structural adjustment is necessary, to support transition.

Table 6.1 (*cont.*)

Reform area	Conclusions and recommendations
Water pollution and water quality: preliminary design considerations	If there is no wider programme of water reform, and specifically no IWRM legislation, consider what high level principles or purposes might be appropriate (e.g., sustainable management or non-deterioration).
	(Alternatively, these may be located in an environmental law. An environmental law may also contain relevant general obligations, e.g., a no-harm rule.)
	Consider whether there are to be combined water use licences including the licensing of discharges; or whether water quality is regulated as part of environmental law. This may determine the regulator.
	Combined water use licences may enable prioritisation of the water agenda and an integrated approach.
	Consider the linkages to other structural regimes, especially conservation law.
	Consider the establishment of a specialist court or tribunal, or specialist prosecutor.
	Consider appropriate mechanisms for public participation, including access to justice/review.
	It is expected that there will be a comprehensive enforcement policy with appropriate powers and sanctions.
	A broad range of sanctions directly available to the regulator is likely to be most effective.
Water pollution and water quality: operational considerations	A clear system of statutory referrals between regulators will ensure good operational links with other pollution control regimes, especially waste management.
	It is expected that there will be a licensing regime for point discharges. A tiered system is recommended.
	It is expected that as a starting point there will be use of emission standards, and the development of quality standards, for common pollutants in the jurisdiction.
	These may be progressively developed as monitoring capacity improves.
	It is feasible to use either guidelines or standards for measuring background environmental quality. Guidelines give flexibility, but may usefully be qualified, for example by a requirement that there must be some improvement in quality even where guidelines are not met, and/or that deviation from the guidelines, and the reasons, be explicitly narrated in some planning documentation.
	More extensive guidelines may supplement a basic set of mandatory parameters.
	Specific quality standards may apply to particular uses, and/or to waters with a conservation designation.
	Take the broadest approach to the management of diffuse pollution, including establishing best practice and appropriate incentives. Consider mandatory provision to back-stop behavioural change and regulatory effort.
Water pollution and water quality: operational considerations for an ecological approach	Taking an ecological approach extends well beyond the scope of traditional pollution control and requires similar data sets to those obtained through an IWRM process; some form of IWRM is therefore a prerequisite.
	Waters will be classified according to their ecological state. There may/may not be a requirement to improve certain classes.
	The classification will entail the setting of standards for ecological flow, and conditions for habitats and biota, including the control of river works, as well as physico-chemical quality standards. These targets/limits can then be reflected in water use authorisations.
	If there are different regulators for different uses, these will require a coordinated approach.
	It will also be necessary to provide for coordination with other strategic regimes, especially land use planning and conservation.
Water services: relationship with water resources law	Water services are a sectoral use of water to which special sectoral rules apply. There are special design considerations around the relationship with the broad water resources framework.
	WSPs will have their own water resource planning functions which should be aligned with and linked to any IWRM processes, with duplication avoided.
	Catchment protection/water safety plans and drought/emergency planning are of particular relevance.

Table 6.1 (*cont.*)

Reform area	Conclusions and recommendations
	WSPs will make both abstractions and discharges, which must be licensed. This may be through the general law, as above, or separate regulation. They are likely to have protected or priority status, at least for abstractions.
	In addition, it is necessary to provide for regulation of prices and of specialist service obligations.
	If there are separate bulk water institutions supplying significant proportions of water services, it is feasible to legislate for such within the same legislation as for resource management and water abstraction. Rural water institutions (e.g., irrigation organisations) may also be legislated for in this way.
	Otherwise, the provision of urban water services is a separate reform agenda and it is not recommended that this should take place at the same time as reform on the broad IWRM agenda or its functional elements.
Water services: preliminary design considerations	Identify levels of unmet need.
	Consider whether the industry should be vertically aggregated, or whether bulk abstraction and treatment, distribution and retail services might be separately operated.
	Consider in principle whether private sector participation will be permitted or encouraged. If so and if disaggregated, consider which elements might be open to private enterprise and which forms of PSP might be appropriate/acceptable.
	Insofar as any service delivery will be provided by the public sector, consider options: national or regional board or agency; local government.
	If local government, there will be implications for business planning and financial regulation. Consider whether local government is the best mechanism for delivery of these services.
	Consider the following:
	Options for regulation of service standards and economic regulation: government department, sectoral regulator, multi-sector regulator, competition authority.
	Consumer representation.
	The extent to which rural users are served differently from urban users: via rural water institutions, and/or in terms of different standards of supply.
	Whether there are any relevant constitutional duties/human rights that might affect the supply of water services; or which duties of supply will complement/implement.
	Appropriate mechanisms for enforcement of duties of supply: public administrative law, and/ or private law; individual action and/or via public authority. A combination of mechanisms is likely to provide most accountability.
	Attitude in principle to subsidy by central government (especially for networks), and to cross-subsidy between categories of users (or for local authorities, across other services).
	Attitude in principle to domestic metering and to availability of disconnections and service restrictions.
	Whether in principle there will be stepped or two-part tariffs, and any minimum entitlement (e.g., 'free' water).
Water services: operational considerations	Establish standards for service provision:
	Universal service obligation; or, identify limitations to service obligations.
	Drinking water quality. (These may be guidelines. There may be fewer parameters in small supply systems.)
	Access and availability: e.g., distance; minimum volume entitlements.
	Pressure and flow.
	Sanitation/wastewater provision. (This may be different in rural and urban areas.)
	Liability for sewer flooding.
	Liability for management of wastewater, and any sludges and/or solid wastes.
	Establish duties with regard to network extension and connections; operation and maintenance; subsidiary powers (e.g., breaking streets, entry to land).

Table 6.1 (*cont.*)

Reform area	Conclusions and recommendations
	Establish mechanisms for business planning and financial auditing. (Delivery by local government is likely to prevent solely sectoral regulation, but there may be oversight regulation by a multi-sectoral agency.)
	Consider how to engage customers in decisions about service levels and costs, or how to require WSPs to do so.
	Establish customer service standards and any compensation mechanisms, or require WSPs to do so, subject to approval by the regulator.
	Establish mechanisms for the investigation of complaints: internal review, complaint to regulator, external review (e.g., ombudsman, court).
	Ensure that all relevant data on WSP performance and costs is publicly available, including contracts and licences, regardless of the business form of the provider.
Water services: conservation	Consider appropriate duties on government, WSPs, regulators (e.g., sustainable use, conservation, efficient use).
	Consider necessary powers to manage drought and water scarcity and provide emergency supply.
	Establish best practices for water efficiency in domestic and commercial use, and for wastewater treatment and reuse. These may be incentivised, or mandatory, or both.
	Consider relationships with planning, building control, plumbing and drainage, and roads authorities.
	Consider mandatory water efficiency measures for newbuild.
	Consider integrating surface/stormwater management into water law.
	Consider whether any duties should be applied to water users (e.g., no waste).
Monitoring and review	The above framework sets out a reform process, which will involve a major legislative programme. It is both unlikely and undesirable that the process should need to be repeated in the medium term. Nonetheless, there should be some adaptability built into the system, so that it can be modified with experience. River basin management is an iterative process, as is an allocation regime, at least in its early stages, and also an ecological approach. Secondary regulation is appropriate, e.g., for detailed rules, and can be relatively easily amended as implementation is monitored and reviewed.
	Consideration should be given by policymakers as to what parameters should be used to measure the effectiveness of the new law, and responsibility designated for their collation and analysis.
	For example, data about dispute settlement, or the degree of compliance with drinking water standards, or the achievement of targets for water quality.
	Stakeholder mechanisms established under the new laws should assist in this regard.

References

International legal instruments

Convention on Access to Information, Public Participation in Decision-making and Access to Justice in Environmental Matters (UN/ECE) (Aarhus) 38 ILM (1999) 517.

Convention on Biological Diversity (CBD) (UN) 31 ILM (1992) 818.

Convention on Economic, Social and Cultural Rights (UN) (1966) UNTS Vol.993 3.

Convention on Elimination of Discrimination against Women (UN) (1979) UNTS Vol.1249 13.

Convention on the Law of the Non-Navigational Uses of International Watercourses (UN) 36 ILM (1997) 700.

Convention on the Rights of the Child (UN) (1989) UNTS Vol.1577 3.

Convention on Wetlands of International Importance (UN) (Ramsar) 11 ILM (1972) 963.

European Convention for the Protection of Human Rights and Fundamental Freedoms (1950) 213 UNTS 221.

Revised Protocol on Shared Watercourses in the SADC Region 40 ILM (2001) 321.

Treaty of Union 22 July 1706.

UN (1972) Declaration of the UN Conference on the Human Environment (Stockholm) A/CONF/48/14/REV.1.

UN (1977) Report of the UN Water Conference (Mar del Plata) E/CONF.70/2.9.

UN (1992) Declaration of the UN Conference on Environment and Development (Rio) A/CONF.151/26.

UN (1992a) Agenda 21: An Agenda for the 21st Century A/Conf. 152/126.

UN (2002) Report of the World Summit on Sustainable Development Incorporating the Johannesburg Declaration and Plan of Implementation (Johannesburg) A/Conf.199/20.

UN (2012) Report of the UN Conference on Sustainable Development (Rio+20) 'The Future We Want' A/CONF.216/L.1.

UN Committee on Economic, Social and Cultural Rights (2002) General Comment No. 15 on the Right to Water E/C.12/2002/11.

UN General Assembly (2000) Millennium Declaration A/RES/55/2.

UN General Assembly (2009) Report of the Independent Expert on the Issue of Human Rights Obligations Related to Access to Safe Drinking Water and Sanitation A/HRC/12/24.

UN General Assembly (2010) Report of the Independent Expert on the Issue of Human Rights Obligations Related to Access to Safe Drinking Water and Sanitation A/HRC/15/31.

UN General Assembly (2010a) GA/10967 available at http://www.un.org/News/Press/docs/2010/ga10967.doc.htm (Accessed 27/03/2014).

UN General Assembly (2013) Report of the Special Rapporteur on the Human Right to Safe Drinking Water and Sanitation, Catarina de Albuquerque A/HRC/24/44.

UN General Assembly Resolution (2010) The Human Right to Water and Sanitation A/64/L.63/Rev.1.

UN Human Rights Council Resolution (2010) Access to Safe Drinking Water and Sanitation A/HRC/15/L.4.

EUROPEAN UNION LEGISLATION

Directive 1976/160/EEC Concerning the Quality of Bathing Water.

Directive 1976/464/EEC on Water Pollution by Dangerous Substances.

Directive 1979/409/EEC on the Conservation of Wild Birds.

Directive 1980/68/EEC on the Protection of Groundwater against Pollution.

Directive 1980/778/EEC on the Quality of Water Intended for Human Consumption.

Directive 86/278/EEC on the Protection of the Environment, and in Particular of the Soil, when Sewage Sludge is used in Agriculture.

Directive 1991/271/EC Concerning Urban Waste Water Treatment.

Directive 1991/676/EEC Concerning the Protection of Waters against Pollution Caused by Nitrates from Agricultural Sources.

Directive 1992/43/EC on the Conservation of Natural Habitats and Wild Flora and Fauna.

Directive 1996/61/EC Concerning Integrated Pollution Prevention and Control.

Directive 1998/83/EC on the Quality of Water Intended for Human Consumption.

Directive 2000/60/EC Establishing a Framework for Community Action in the Field of Water Policy.

Directive 2000/60/EC Framework Directive for Water Policy.

Directive 2001/42/EC on the Assessment of the Effects of Plans and Programmes on the Environment.

Directive 2003/35/EC Providing for Public Participation in Respect of the Drawing up of Certain Plans and Programmes Relating to the Environment.

Directive 2003/4/EC on Public Access to Environmental Information.

Directive 2006/11/EC on Pollution Caused by Certain Dangerous Substances Discharged into the Aquatic Environment of the Community.

Directive 2006/113/EC on the Quality Required of Shellfish Waters.

Directive 2006/118/EC on the Protection of Groundwater against Pollution and Deterioration.

Directive 2006/44/EC on the Quality of Fresh Waters Needing Protection or Improvement in Order to Support Fish Life.

Directive 2006/7/EC Concerning the Management of Bathing Water Quality.

Directive 2007/60/EC on the Assessment and Management of Flood Risks.

Directive 2008/1/EC Concerning Integrated Pollution Prevention and Control.

Directive 2008/56/EC Establishing a Framework for Community Action in the Field of Marine Environmental Policy (Marine Strategy Framework Directive).

Directive 2008/105/EC on Environmental Quality Standards in the Field of Water Policy.

Directive 2009/128/EC Establishing a Framework for Community Action to Achieve the Sustainable Use of Pesticides.

Directive 2009/147/EC on the Conservation of Wild Birds.

Directive 2010/75/EU on Industrial Emissions (Integrated Pollution Prevention and Control).

Directive 2011/92/EU on the Assessment of the Effects of Certain Public and Private Projects on the Environment.

Directive 2013/39/EU as regards Priority Substances in the Field of Water Policy.

Recommendation (2002/413/EC) Concerning the Implementation of Integrated Coastal Zone Management in Europe.

Regulation EC/1107/2009 Concerning the Placing of Plant Protection Products on the Market.

Regulation EC/1907/2006 Concerning the Registration, Evaluation, Authorisation and Restriction of Chemicals.

Regulation EC/73/2009 Establishing Common Rules for Direct Support Schemes for Farmers under the Common Agricultural Policy and Establishing Certain Support Schemes for Farmers.

National legal instruments

AUSTRALIA

Constitution of Australia Act 1900 63 and 64 Vict. c.12.

Environmental Protection and Biodiversity Conservation (Cwlth) Act 1999 No.91.

Great Barrier Reef and Marine Parks (Cwlth) Act 1975 No.85.

Water (Cwlth) Act 2007 No.137.

Water Amendment (Cwlth) Act 2008 No.139.

Water Efficiency Labelling and Standards (Cwlth) Act 2005 No.4.

Intergovernmental Agreement (2013) between the Commonwealth and States on Implementing Water Reform in the Murray—Darling Basin available at http://www.coag.gov.au/node/506 (Accessed 16/04/2014).

Commonwealth of Australia Water Act 2007 Basin Plan 2012 F2012L02240.

QUEENSLAND

City of Brisbane (Qld) Act 2010 No.23.

Coastal Protection and Management (Qld) Act 1995 No.41.

Constitution of Queensland Act 2001 No.80.

Disaster Management (Qld) Act 2003 No.93.

Environment Protection (Qld) Act 1994 No.62.

Environmental Offsets (Qld) Bill 2014.

Environmental Protection (Greentape Reduction) and Other Legislation Amendment (Qld) Act 2012 No.16.

Government Owned Corporations (Qld) Act 1993 No.38.

Land Titles (Qld) Act 1994 No.11.

Land, Water and Other Legislation Amendment (Qld) Act 2013 No.23.

Local Government (Qld) Act 2009 No.17.

Marine Parks (Qld) Act 2004 No.31.

New South Wales–Queensland Border Rivers (Qld) Act 1946 11 Geo.6 No.16.

Ombudsman (Qld) Act 2001 No.73.

Plumbing and Drainage and Other Legislation Amendment (Qld) Act 2005 No.39.

Public Health (Qld) Act 2005 No.48.

Queensland Competition Authority (Qld) Act 1997 No.25.

Right to Information (Qld) Act 2009 No.13.

South East Queensland Water (Distribution and Retail Restructuring) (Qld) Act 2009 No.46.

South East Queensland Water (Distribution and Retail Restructuring) (Qld) Act 2012 No.1.

South East Queensland Water (Restructuring) (Qld) Act 2007 No. 58.

South East Queensland Water (Restructuring) and Other Legislation Amendment (Qld) Act 2012 No.39.

State Development and Public Works (Qld) Act 1971 No.55.

Sustainable Planning (Qld) Act 2009 No.36.

Waste Reduction and Recycling (Qld) Act 2011 No.31.

Water (Commonwealth Powers) (Qld) Act 2008 No. 58.

Water (Qld) Act 2000 No.34.

Water Amendment (Qld) Act 2006 No.23.

Water and Other Legislation Amendment (Qld) Act 2010 No.53.

Water and Other Legislation Amendment (Qld) Act 2011 No. 40.

Water Efficiency Labelling and Standards (Qld) Act 2005 No.69.

Water Fluoridation (Qld) Act 2008 No.12.

Water Supply (Safety and Reliability) (Qld) Act 2008 No.34.

Wild Rivers (Qld) Act 2005 No.42.

Regulations

Environmental Protection (Water) Amendment Policy (No.2) 2013 SL No.272.

Environmental Protection (Water) Policy 1997 SL No.136.

Environmental Protection (Water) Policy 2009 SL No.178.

Environmental Protection Regulation 2008 SL No.370.

Local Government Regulation 2012 SL No.236.

Public Health Regulation 2005 SL No.281.

Queensland Competition Regulation 2007 SL No.207.

Sustainable Planning Regulation 2009 SL No.280.

Water Amendment Policy (No.2) 2013 SL No.272.

Water Regulation 2002 SL No.70.

Water (Transitional) Regulation 2012 SL No.242.

Water Resource (Condamine and Balonne) Plan 2004 SL No.151.

Water Resource (Gold Coast) Plan 2006 SL No.321.

SOUTH AFRICA

Constitution of South Africa Act 1996 No.108.

Disaster Management Act 2002 No.57.

Distribution of Revenue Act 2013 No.2.

Integrated Coastal Management Act 2008 No.24.

Marine Living Resources Act 1998 No.18.

Mineral and Petroleum Resources Development Act 2002 No.28.

Municipal Finance Management Act 2003 No.56.

Municipal Systems Act 2000 No.32.

Municipal Systems Act 2003 No.44.

National Environmental Management (Waste Management) Act 2008 No.59.

National Environmental Management Act 1998 No.107.

National Environmental Management Amendment Act 2003 No.46.

National Environmental Management Amendment Act 2004 No.8.

National Environmental Management Environmental Laws Amendment Act 2008 No.44.

National Health Act 2003 No.61.

National Water Act 1998 No.36.

Promotion of Access to Information Act 2000 Act No.2.

Public Finance Management Act 1999 No.1.

Public Finance Management Act 1999 No.29.

South African National Water Resources Infrastructure Agency Limited Bill 2008 B36–2008.

Water Services Act 1997 No.108.

Regulations

Establishment of the Water Management Areas as a Component of the National Water Resource Strategy: General Notice No.1160 of 1999.

General Authorisation in Terms of Section 39 of the National Water Act: General Notice No.399 of 2004.

General Authorisation in Terms of Section 39 of the National Water Act: General Notice No.1199 of 2009.

General Authorisation in Terms of Section 39 of the National Water Act: General Notice No.665 of 2013.

National Water Policy Review General Notice No.888 of 2013.

Norms and Tariffs in Respect of Water Services Regulation No.652 of 2001.

Regulations for the Establishment of a Water Resource Classification System No.810 of 2010.

Regulations in Terms of the National Environmental Management Act Nos.385, 386, 387 of 2006.

Regulations on Financial Assistance to Resource Poor Farmers No.1036 of 2007.

Regulations Relating to Compulsory National Standards and Measures to Conserve Water No.509 of 2001.

Regulations Requiring that a Water Use be Registered No.1352 of 1999.

Water Services Provider Contract Regulations No.980 of 2002.

Draft General Authorisation in Terms of Section 39 of the National Water Act General Notice No.288 of 2012.

Draft Technical Regulations for Petroleum Exploration and Exploitation General Notice 1032 of 2013.

UNITED KINGDOM

Competition Act 1998 c.41.
Control of Pollution Act 1974 c.40.
Electricity Act 1989 c.29.
Enterprise Act 2002 c.40.
Enterprise and Regulatory Reform Act 2013 c.24.
Environment Act 1995 c.25.
Environmental Protection Act 1990 c.43.
European Communities Act 1972 c.68.
Floods and Water Management Act 2010 c.29.
Human Rights Act 1998 c.42.
Marine and Coastal Access Act 2009 c.23.
Reservoirs Act 1975 c.23.
Scotland Act 1998 c.46.
Scotland Act 2012 c.11.
Utilities Act 2000 c.27.
Water (Fluoridation) Act 1985 c.63.
Water Act 1989 c.15.
Water Act 2003 c.37.
Water Bill 2013 Bill No.82.

Regulations

Marine Strategy Regulations SI 2010/1627.
Water Environment (Water Framework Directive) (Northumbria RBD) Regulations SI 2003/3245.
Water Environment (Water Framework Directive) (Solway Tweed RBD) Regulations SI 2004/99.

ENGLAND

Highways Act 1980 c.66.
Tribunals, Courts and Enforcement Act 2007 c.15.
Water Industry Act 1991 c.36.
Water Resources Act 1963 c.38.
Water Resources Act 1991 c.57.

Regulations

Civil Procedure (Amendment) Rules SI 2013/262.
Drought Plans Regulations SI 2005/1905.

Environmental Assessment of Plans and Programmes Regulations SI 2004/1663.
Environmental Civil Sanctions (England) Order SI 2010/1157.
Environmental Information Regulations SI 2004/3391.
Environmental Permitting (England and Wales) Regulations SI 2010/675.
Flood Risk Regulations SI 2009/3042.
Nitrate Pollution Prevention Regulations SI 2008/2349.
Town and Country Planning (Environmental Impact Assessment) Regulations SI 2011/1824.
Urban Waste Water Treatment (England and Wales) Regulations SI 1994/2841.
Water Environment (Water Framework Directive) (England and Wales) Regulations SI 2003/3242.
Water Industry (Charges) (Vulnerable Groups) Regulations SI 1999/3441.
Water Industry (Prescribed Conditions) Regulations SI 1999/3422, SI 2007/2457.
Water Industry (Schemes for Adoption of Private Sewers) Regulations SI 2011/1566.
Water Resources (Abstraction and Impounding) Regulations SI 2006/641.
Water Resources (Control of Pollution) (Silage, Slurry and Agricultural Fuel Oil) (England) Regulations SI 2010/639, 2010/1091.
Water Supply and Sewerage Services (Customer Service Standards) Regulations SI 1989/1159.
Water Supply (Water Fittings) Regulations SI 1999/1148.
Water Use (Temporary Bans) Order SI 2010/2231.

SCOTLAND

Act of Union with England 1707 c.7.
Environmental Assessment (Scotland) Act 2005 asp.15.
Flood Risk Management (Scotland) Act 2009 asp.6.
Freedom of Information (Scotland) Act 2002 asp.13.
Land Reform (Scotland) Act 2003 asp.2.
Local Government (Scotland) Act 1994 c.39.
Marine (Scotland) Act 2010 asp.5.
Natural Heritage (Scotland) Act 1991 c.28.
Public Services Reform (Scotland) Act 2010 asp.8.
Reservoirs (Scotland) Act 2011 asp.9.
Roads (Scotland) Act 1984 c.54.
Scottish Public Services Ombudsman Act 2002 asp.11.
Sewerage (Scotland) Act 1968 c.47.
Water (Scotland) Act 1980 c.45.
Water Environment and Water Services (Scotland) Act 2003 asp.3.
Water Industry (Scotland) Act 2002 asp.3.
Water Resources (Scotland) Act 2013 asp.5.

Regulatory Reform (Scotland) Bill 2013 SP Bill 26 Session 4.

Regulations

Act of Sederunt (Rules of the Court of Session Amendment) (Protective Expenses Orders in Environmental Appeals and Judicial Reviews) SSI 2013/81.

Action Programme for Nitrate Vulnerable Zones (Scotland) Regulations SSI 2008/298.

Environmental Information (Scotland) Regulations SSI 2004/520.

Pollution Prevention and Control (Scotland) Regulations SSI 2012/360.

Private Water Supplies (Scotland) Regulations SSI 2006/209.

Silage, Slurry and Agricultural Fuel Oil (Scotland) Regulations SSI 2003/531.

Town and Country Planning (Environmental Impact Assessment) (Scotland) Regulations SSI 2011/139.

Urban Waste Water Treatment (Scotland) Regulations SI 1994/2842.

Waste Management Licensing (Scotland) Regulations SSI 2011/228.

Water Environment (Consequential and Savings Provisions) (Scotland) Order SSI 2006/181.

Water Environment (Controlled Activities) (Scotland) Regulations SSI 2005/348.

Water Environment (Controlled Activities) (Scotland) Regulations SSI 2011/209.

Water Environment (Diffuse Pollution) (Scotland) Regulations SSI 2008/54.

Water Environment (Oil Storage) (Scotland) Regulations SSI 2006/133.

Water Environment (Register of Protected Areas) (Scotland) Regulations SSI 2004/516.

Water Environment (Relevant Enactments and Designation of Responsible Authorities and Functions) (Scotland) Order SSI 2011/368.

Water Environment and Water Services (Scotland) Act 2003 (Consequential Provisions and Modifications) Order SI No.2006/1054.

Water Services etc. (Scotland) Act 2005 (Consequential Provisions and Modifications) Order SI 2005/3172.

Water Supply (Water Quality) (Scotland) Regulations SI 1990/119.

Water Byelaws (Scotland) 2004 available at http://www.scottishwater.co.uk/business/our-services/compliance/water-byelaws/water-byelaws-documents (Accessed 27/03/2014).

LIST OF CASES

Aguas del Tunari S.A. v Republic of Bolivia (ICSID Case No. ARB/02/3).

Albion Water Ltd v Dwr Cymru Cyfyngedig [2013] CAT 16.

Albion Water Ltd v Water Services Regulation Authority (Dwr Cymru/Shotton Paper) [2006] CAT 23, [2008] EWCA Civ 536, [2008] UKCLR 457.

Albion Water Ltd v Water Services Regulation Authority and Dwr Cymru Cyfyngedig, United Utilities Water PLC intervening [2008] CAT 31.

Alphacell v Woodward [1972] AC 824.

AXA General Insurance Limited and Others (Appellants) v The Lord Advocate and Others (Respondents) (Scotland) [2011] UKSC 46.

Beja and Others v Premier of the Western Cape and Others (2011) Case No. 21332/10.

Biwater Gauff (Tanzania) Limited v United Republic of Tanzania (ICSID Case No. ARB/05/22) (Award: 24 July 2008).

Catscratch Ltd No.2 v Glasgow City Licensing Board 2002 SLT 503.

Colquhoun's Trustees v Orr Ewing (1877) 4R HL 116.

Compañía de Aguas del Aconquija S.A. and Vivendi Universal S.A. v Argentine Republic (ICSID Case No. ARB/97/3) (Award: 20 August 2007).

Crown Estates Cssnrs. v Fairlie Yacht Slip Ltd. 1979 SC 156.

Empress Car Co Ltd. v National Rivers Authority [1999] 2 AC 22.

European Commission v Germany C-525/12 (Action brought 19 November 2012).

European Commission v UK [1992] ECR I-6103 C-337/89.

European Commission v UK [1999] ECR I-2023 C-340/96.

European Commission v UK [2013], [2014] ECR 0000 C-530/11.

Fish Legal (and another) v The Information Commissioner, United Utilities, Yorkshire Water and Southern Water Case C-279/12.

Fish Legal v IC [2012] UKUT 177 (AAC).

Fredin v Sweden (1991) 13 EHRR 784.

Harmony Gold Mining Company Ltd v Regional Director: Free State Department of Water Affairs (971/12) [2013] ZASCA 206.

Jacobsson v Sweden (1990) 12 EHRR.

Lord Advocate v Clyde Navigation Trustees (1891) 19R 174.

Manquele v Durban Transitional Metropolitan Council Case No. 2036/2000.

Marcic v Thames Water PLC [2004] 1 All ER 135.

Mazibuko v City of Johannesburg (39/09) [2009] ZACC 28.

McColl v Strathclyde Regional Council 1983 SC 225.

Mellacher v Austria (1990) 12 EHRR 391.

Minister for the Environment and Heritage v Queensland Conservation Council Inc. [2004] FCAFC 190.

Nokotanya and Others v Ekurhuleni Metropolitan Municipality and Others (31/09) [2009] ZACC 33.

Price v Cromack [1975] 1 WLR 988.

Pyle v Gilbert (1980) 245 Ga. 403, 265 S.E.2d 584.

R (Edwards & Pallikaropoulos) v. Environment Agency and Others [2012], [2013] C-260/11.

R v Director of Water Services ex p Lancashire County Council and Others [1999] EnvLR 114.

R v Secretary of State for the Environment ex parte Friends of the Earth [1995] EnvLR 11.

Shetland Salmon Farmers Association v Crown Estates Cssnrs 1991 SLT 166.

Smartsource v IC and a Group of 19 additional water companies [2010] UKUT 415 (AAC).

The Scottish Environment Protection Agency and Others (Reclaimers) [2013] CSIH 108.

Walton (Appellant) v The Scottish Ministers (Respondent) (Scotland) [2012] UKSC 44.

Wills Trustees v Cairngorm Canoeing Club 1976 SLT 162.

Journal articles and conference papers

Abrams, L. J. (1996) Policy Development in the Water Sector: The South African Experience. Presented at the Cranfield International Water Policy Conference, Cranfield University, Bedford, UK, September 1996, available at http://www.africanwater.org/sapolicy.htm (Accessed 26/03/2014).

Allan, A. (2005) The Legal Context of Water User Associations in Nepal. *Denver Water Law Review* 8(2) 547–601.

Anderson, A. J. (2005) Engaging Disadvantaged Communities: Lessons from the Inkomati CMA Establishment Process. Presented at International Workshop on African Water Laws, Johannesburg, 2005, available at http://www.nri.org/waterlaw/workshop.htm (Accessed 19/03/2014).

Arnstein, S. (1969) A Ladder of Citizen Participation. *American Institute of Planners Journal* 35 216–224.

Bakker, K. (2003a) From Public to Private to Mutual? Restructuring Water Supply Governance in England and Wales. *Geoforum* [Amsterdam] 34(3) 359–374, available at http://www.geog.ubc.ca/~bakker/Publications/ (Accessed 27/03/2014).

Baumfield, V. (2012) The Allconnex Water Debacle: Lessons in Devising Better Governance Mechanisms for Government Business Enterprises. *Bond Law Review* 24.2 1–63.

Biswas, A. (2004) Integrated Water Resources Management: A Reassessment. *Water International* 29(2) 248–256.

Biswas, A. (2008) Integrated Water Resources Management: Is It Working? *International Journal of Water Resources Development* 24(1) 5–22.

Biswas, A. and Tortajada, C. (2010) Future Water Governance: Problems and Perspectives. *International Journal of Water Resources Development* 26(2) 129–139.

Bonn Nexus (2011) The Water, Energy and Food Security Nexus: Solutions for a Green Economy. Policy Recommendations available at http://www.water-energy-food.org/en/whats_the_nexus/messages_policy_recommendations.html (Accessed 17/03/2014).

Burchi, S. (2005) The Interface between Customary and Statutory Water Rights – a Statutory Perspective. Presented at the International Workshop on African Water Laws, Johannesburg, 26–28 January 2005, available at http://projects.nri.org/waterlaw/workshop.htm (Accessed 24/03/2014).

Calatrava, J. and Garrido, A. (2006) Difficulties in Adopting Formal Water Trading Rules within Users' Associations. *Journal of Economic Issues* XL (1) 27.

Carter, J. and Howe, J. (2006) The Water Framework Directive and the Strategic Environmental Assessment Directive: Exploring the Linkages. *Environmental Impact Assessment Review* 26(3) 287–300.

Clark, B. (2002) Migratory Things on Land: Property Rights and a Law of Capture. *Electronic Journal of Comparative Law* 6.3, available at http://www.ejcl.org/63/abs63-3.html (Accessed 24/03/2014).

Clark, B. (2006) Water Law in Scotland: The Water Environment and Water Services (Scotland) Act 2003 and the European Convention on Human Rights. *Edinburgh Law Review* 10 60–101.

Connell, D. (2011) Water Reform and the Federal System in the Murray–Darling Basin. *Water Resources Management* 25 3993–4003.

Dellapenna, J. (2000) The Importance of Getting Names Right: The Myth of Markets for Water. *William & Mary Environmental Law and Policy Review* 25 317–377.

Eaton, T. (2013) Science-based Decision-making on Complex Issues: Marcellus Shale Gas Hydrofracking and New York City Water Supply. *Science of the Total Environment* 461–462 158–169 available at http://dx.doi.org/10.1016/j.scitotenv.2013.04.093 (Accessed 26/03/2014).

Ebbesson, J. (2013) The Aarhus Convention Access to Justice and Compliance by the UK. *Journal of Environmental Law and Management* 25(2–3) 56–60.

Garrick, D. *et al.* (2012) Environmental Water Governance in Federal Rivers: Opportunities and Limits for Subsidiarity in Australia's Murray–Darling River. *Water Policy* 14 915–936.

Gleick, P. (1996) Basic Water Requirements for Human Activities Meeting Basic Needs. *Water International* 21 (1996) 83–92.

Godden, L. (2005) Water Law Reform in Australia and South Africa: Sustainability Efficiency and Social Justice. *Journal of Environmental Law* 17 181.

Grafton, R. and Jiang, Q. (2011) Economic Effects of Water Recovery on Irrigated Agriculture in the Murray-Darling Basin. *The Australian Journal of Agricultural and Resource Economics* 55 487–499.

Gunningham, N. and Sinclair, D. (2007) Policy Instrument Choice and Diffuse Source Pollution. *Journal of Environmental Law* 17(1) 51.

Hardin, G. (1968) The Tragedy of the Commons. *Science* 162 1243–1248.

Hendry, S. (2003) Scotland's Water: Safe Clean Affordable Public. *Natural Resources Journal* 43 491–517.

Hendry, S. (2013) Private Rights and Public Responsibilities: Recent Developments in Scots Water Law Policy Brief, The Foundation for Law, Justice and Society, in association with the Centre for Socio-Legal Studies and Wolfson College, University of Oxford, available at http://www.fljs.org/sites/www.fljs.org/files/publications/Hendry.pdf (Accessed 24/03/2014).

Hendry, S. and Reeves, A. (2012) The Regulation of Diffuse Pollution in the European Union: Science, Governance and Water Resource Management. *International Journal of Rural Law and Policy, North America*, available at http://epress.lib.uts.edu.au/journals/index.php/ijrlp/article/view/2699 (Accessed 24/03/2014).

Howarth, W. (2005) Editorial. *Journal of Water Law* 16 3–4.

Kaika, M. (2003) The Water Framework Directive: A New Directive for a Changing Social, Political and Economic European Framework. *European Planning Studies* 11(3) 299–316.

Kallis, G. and Nijkamp, P. (2000) Evolution of EU Water Policy: A Critical Assessment and a Hopeful Perspective. *Journal of Environmental Law and Policy* 3 301–355.

Littlechild, S. (1988) Economic Regulation of the Privatised Water Authorities and Some Further Reflections. *Oxford Review of Economic Policy* 4.

Maltby, E. and Acreman, M. (2011) Ecosystem Services of Wetlands: Pathfinder for a New Paradigm. *Hydrological Sciences Journal* 56(8) 1341–1359.

McCaffrey, S. (1992) A Human Right to Water: Domestic and International Implications. *Georgetown International Environmental Law Review* 5(1), 8.

Muller, M. (2008) Free Basic Water A Sustainable Instrument for a Sustainable Future in South Africa. *Environment and Urbanization* 20(1) 67–87.

Muller, M. (2010) Fit for Purpose: Taking Integrated Water Resource Management Back to Basics. *Irrigation and Drainage Systems* 24 161–175.

Ostrom, E. *et al.* (1999) Revisiting the Commons: Local Lessons, Global Challenges. *Science* 284(5412) 278–282.

Parker, S. (2003) Water Trading in Queensland. Paper presented at the 8th Annual National Water Conference, Sydney, 24–25 June 2003 (copy on file with the author).

Pegram, G. and Bofilatos, E. (2005) Considerations on the Composition of CMA Governing Boards to Achieve Representation. Presented at International Workshop on African Water Laws, Johannesburg, 2005, available at http://www.nri.org/waterlaw/workshop.htm (Accessed 19/03/2014).

Pool, R. (2013) Can a Complex Canal System Solve Britain's Water Woes? *Engineering and Technology Magazine* 8(8) 1, available at http://eandt.theiet.org/magazine/2013/08/britains-root-canal.cfm (Accessed 24/03/2014).

Scott, A. and Coustalin, G. (1995) The Evolution of Water Rights. *Natural Resources Journal* 35 821–979.

Stern, J. and Holder, S. (1999) Regulatory Governance: Criteria for Assessing the Performance of Regulatory Systems: An Application to Infrastructure Industries in the Developing Countries of Asia. *Utilities Policy* 8 33–50.

Stoker, G. (1998) Governance as Theory: Five Propositions. *International Social Science Journal* 155 17–27.

Tan, P. (2002) The Changing Concepts of Property in Surface Water Resources in Australia. *Journal of Water Law* 13 269–275.

Tremblay, H. (2011) A Clash of Paradigms in the Water Sector? Tensions and Synergies between Integrated Water Resources Management and the Human Rights-Based Approach to Development. *Natural Resources Journal* 51 307–356.

Wallace, P. and Ison, R. (2011) Appreciating Institutional Complexity in Water Governance Dynamics: A Case from the Murray-Darling Basin, Australia. *Water Resources Management* 25 4081–4097.

Waska, M. and Gardner, A. (2012) *Using Regulation to Tackle the Challenge of Diffuse Water Pollution and Its Impact on the Great Barrier Reef*. University of Western Australia Faculty of Law Research Paper No. 2013-23.

Watson, N. *et al.* (2007) Critical Perspectives on Integrated Water Management (Editorial). *Geographical Journal* 173(4) 297–299.

Wiel, S. (1919) Waters: American Law and French Authority. *Harvard Law Review* XXXIII(2) 133.

Wittwer, G. (2011) Confusing Policy and Catastrophe: Buybacks and Drought in the Murray–Darling Basin. *Economic Society of Australia Economic Papers*, 30(3) 289–295.

Books

Bakker, K. (2003) *An Uncooperative Commodity: Privatising Water in England and Wales*. Oxford: Oxford University Press.

Ball, P. (2002) *H₂O: A Biography of Water*. London: Phoenix Press.

Bauer, C. (2004) *Siren Song: Chilean Water Law as a Model for International Reform*. Washington, DC: Resources for the Future.

Bell, S. *et al.* (2013) *Environmental Law*. Oxford: Oxford University Press.

Bennett, J. (Ed.) (2005) *The Evolution of Markets for Water*. Cheltenham: Edward Elgar.

Biswas, A. and Tortejada, C. (Eds.) (2009) *Impacts of Megaconferences on the Water Sector*. Mexico Third World Centre for Water Management/Springer.

Bruns, B., Ringler, C. and Meinzen-Dick, R. (Eds.) (2005) *Water Rights Reform: Lessons for Institutional Design*. Washington, DC: IFPRI.

Caponera, D. (1992) *Principles of Water Law and Administration*. Amsterdam: Balkema.

Chenoweth, J. and Bird, J. (Eds.) (2006) *The Business of Water and Sustainable Development*. Sheffield: Greenleaf Publishing.

Connell, D. and Grafton, R. Q. (Eds.) (2011) *Basin Futures: Water Reform in the Murray-Darling Basin*. Canberra: ANU E-Press.

Cullet, P. *et al.* (Eds.) (2010) *Water Law for the Twenty-First Century*. Abingdon: Routledge.

De Waal, C. (2005) *On Pragmatism*. Belmont, CA: Thomson/Wadsworth.

Dellapenna, J. and Gupta, J. (Eds.) (2008) *The Evolution of the Law and Politics of Water*. Springer.

Delmon, J. (2000) *BOO/BOT Projects A Commercial and Contractual Guide*. London: Sweet & Maxwell.

Delmon, J. (2001) *Water Projects: A Commercial and Contractual Guide*. The Hague: Kluwer Law International.

Emoto, M. (2004) *The Hidden Messages in Water*. Hillsboro, OR: Beyond Words Publishing.

Ferguson, J. (1907) *The Law of Water and Water Rights in Scotland*. Edinburgh: W. Green & Son.

Finger, M. and Allouche, J. (2002) *Water Privatisation*. London: Spon Press.

Fisher, D. (2000) *Water Law*. Sydney: The Law Book Company.

Freeman, M. D. A. (2001) *Lloyd's Introduction to Jurisprudence*. London: Sweet & Maxwell.

Fuggle, R. and Rabie, M. (Eds.) (1992) *Environmental Management in South Africa*. Cape Town: Juta & Co.

Garrido, A. and Ramon Llamas, M. (Eds.) (2009) *Water Policy in Spain*. Leiden: CRC Press/Balkema.

Getzler, J. (2005) *A History of Water Rights at Common Law*. Oxford: Oxford University Press.

Gunningham, N. and Grabosky, P. (1998) *Smart Regulation*. Oxford: Oxford University Press.

Hawkins, K. (1984) *Environment and Enforcement: Regulation and the Social Definition of Pollution*. Oxford: Oxford University Press.

Hodgson, S. (2006) *Modern Water Rights: Theory and Practice*. Rome: FAO Legislative Series 92.

Hu, D. (2006) *Water Rights: An International and Comparative Study*. London: IWA Publishing.

Kenney, D. (Ed.) *In Search of Sustainable Water Management*. Cheltenham: Edward Elgar.

Lee, R. W. (1956) *The Elements of Roman Law*. Edinburgh: Sweet & Maxwell.

Lenton, R. and Muller, M. (2009) *Integrated Water Resources Management in Practice*. London: Earthscan.

Magsig, B.-O. (2013) International Water Security and the Quest for Common Security (Unpublished PhD thesis, University of Dundee).

Malacek, S. (2013) A Legal Framework for Local Private Sector Participation in Urban Water Services (Unpublished PhD thesis, University of Dundee).

McDonald, D. and Ruiters, G. (Eds.) (2005) *The Age of Commodity: Water Privatisation in Southern Africa*. London: Earthscan.

McManus, F. (Ed.) (2007) *Environmental Law*. Edinburgh: SULI/W. Green & Son.

Merret, S. (2005) *The Price of Water*. London: IWA Publishing.

Ogus, A. (1994) *Regulation: Legal Form and Economic Theory*. Oxford: Clarendon Press.

Ostfeld, A. and Tyson, J. (Eds.) (2005) *River Basin Restoration and Management*. London: IWA Publishing.

Pidwirny, M. (2006) *Fundamentals of Physical Geography*, 2nd Edition available at http://www.physicalgeography.net/fundamentals/8o.html (Accessed 17/03/2014).

Reid, K. and Zimmerman, R. (Eds.) (2000) *A History of Private Law in Scotland Vol. I*. Oxford: Oxford University Press.

Ruhl, J. B., Kraft, S. and Lant, C. (2007) *The Law and Policy of Ecosystem Services*. Washington, DC: Island Press.

Salman, S. and Bradlow, D. (2006) *Regulatory Frameworks for Water Resources Management*. Washington, DC: World Bank.

Samuel, G. (2003) *Epistemiology and Method in Law*. Aldershot: Ashgate.

Tarlock, A. D. *et al.* (2009) *Water Resource Management: A Casebook in Law and Public Policy* 6th Edition. New York: Foundation Press.

Tietenberg, T. (2007) *Environmental Economics and Policy* 5th Edition. Boston: Pearson.

Van Koppen, B., Giordano, M. and Butterworth, J. (Eds.) (2007) *Community-Based Water Law and Water Resource Management Reform in Developing Countries*. Wallingford: CAB International.

Vickers, J. and Yarrow, G. (1988) *Privatisation: An Economic Analysis*. Cambridge, MA: MIT Press.

Winkler, I. (2012) *The Human Right to Water*. Oxford: Hart Publishing.

World Commission on Dams (2000) *Dams and Development: A New Framework for Decision-Making*. London: Earthscan.

Policy documents and reports

AFROL News (2004) South Africa Sets Up New Environmental Court (online) 24 February available at http://www.afrol.com/articles/11360 (Accessed 24/03/2014).

Algotsson, E. and Murombo, T. (2009) *Water Supply and Sanitation in South Africa: Environmental Rights and Municipal Accountability* available at http://cer.org.za/wp-content/uploads/2011/11/LHR-DBSA_Water_Report.pdf (Accessed 27/03/2014).

Allan, A. *et al.* (2012) *Legal and Institutional Analysis, Public Participation and Stakeholder Involvement*, Report No. D6.3 EU GENESIS Project available at http://www.bioforsk.no/ikbViewer/page/prosjekt/tema?p_dimension_id=16858&p_menu_id=16904&p_sub_id=16859&p_dim2=23505 (Accessed 24/03/2014).

AMCOW (2011) *Water Supply and Sanitation in South Africa* available at http://www.wsp.org/sites/wsp.org/files/publications/CSO-SouthAfrica.pdf (Accessed 26/03/2014).

Anglian Water (2014) *Draft 2014 Water Resources Management Plan* available at http://www.anglianwater.co.uk/environment/water-resources/resource-management/ (Accessed 19/03/2014).

Angling Trust (2011) *Campaigners Halt Judicial Review* available at http://www.anglingtrust.net/news.asp?section=29&itemid=758 (Accessed 19/03/2014).

ANZECC/ARMCANZ (2000) *Australian and New Zealand Guidelines for Fresh and Marine Water Quality Volume 1* and related documents available at http://www.environment.gov.au/topics/water/water-quality/national-water-quality-management-strategy (Accessed 24/03/2014).

ANZECC/ARMCANZ (2004) *Guidelines for Sewerage Systems: Biosolids Management* available at http://www.biosolids.com.au/aust-nz-guidelines.php (Accessed 27/03/2014).

APSC (2007) *Tackling Wicked Problems: A Public Policy Perspective* available at http://www.apsc.gov.au/__data/assets/pdf_file/0005/6386/wicked-problems.pdf (Accessed 09/04/2013).

Asian Development Bank (1995) *Governance: Sound Development Management* available at http://www.adb.org/documents/governance-sound-development-management (Accessed 19/03/2014).

Associated Programme on Flood Management (2006) *Legal and Institutional Aspects of Integrated Flood Management* available at http://www.apfm.info/pdf/ifm_legal_aspects.pdf (Accessed 19/03/2014).

Auditor-General (2012) *Consolidated General Report on the Audit Outcomes of Local Government 2011–12* available at http://www.agsa.co.za/Documents/Auditreports/MFMA20112012.aspx (Accessed 27/03/2014).

Australian Government/Murray–Darling Basin Authority (2012) *The Socio-Economic Implications of the Proposed Basin Plan* available at http://www.mdba.gov.au/what-we-do/basin-plan/development/steps-in-the-development-of-the-basin-plan (Accessed 20/03/2014).

Breede-Overberg CMA (2010) *Breede-Overberg Catchment Management Strategy (Draft)* available at http://www.bocma.co.za/documents.html (Accessed 19/03/2014).

Building Standards for Scotland Technical Handbook (2013) available at http://www.scotland.gov.uk/Topics/Built-Environment/Building/Building-standards/techbooks/techhandbooks (Accessed 27/03/2014).

COAG (1995) *National Competition Policy and Related Reforms Agreement* and related documents available at http://ncp.ncc.gov.au/pages/about (Accessed 24/03/2014).

Coetzee, H. *et al.* (2010) *Report to the Inter-Ministerial Committee on Acid Mine Drainage: Mine Water Management in the Witwatersrand Gold Fields with Special Emphasis on Acid Mine Drainage* available at http://www.dwaf.gov.za/Documents/ACIDReport.pdf (Accessed 24/03/2014).

CoGTA (2013) *Annual Report 2012/13* available at http://www.kzncogta.gov.za/Portals/0/Documents/reports/2012/1319027%20COGTA%20Low%20Res.pdf (Accessed 27/03/2014).

COHRE (2003) *Legal Resources for the Right to Water* available at http://www.cohre.org/sites/default/files/sources_8_-_legal_resources_for_the_right_to_water_october_2003.pdf (Accessed 27/03/2014).

COHRE (2008) *Water Services Fault Lines: An Assessment of South Africa's Water and Sanitation Provision across 15 Municipalities* available at http://www.cohre.org/sites/default/files/water_services_fault_lines_sa_nov08.pdf (Accessed 27/03/2014).

Condamine Alliance (2010) *Natural Resource Management Plan* available at http://www.condaminealliance.com.au/general-publications (Accessed 20/03/2014).

Cottingham, R. *et al.* (2010) *Managing Diffuse Water Pollution in South East Queensland: An Analysis of the Role of the Healthy Waterways Partnership* available at http://www.watercentre.org/resources/publications/attachments/Diffuse_Pollution_SEQ_IWC_20110412_Cottingham_Delfau_Garde.pdf (Accessed 24/03/2014).

Council of Australian Governments (1992) *The National Strategy for Ecologically Sustainable Development* available at http://www.environment.gov.au/resource/national-strategy-ecologically-sustainable-development (Accessed 20/03/2014).

Council of Australian Governments (1994) *Water Reform Framework* available at http://www.environment.gov.au/resource/council-australian-governments-water-reform-framework (Accessed 20/03/2014).

Council of Australian Governments (2004) *National Water Initiative* (NWI) available at http://nwc.gov.au/nwi (Accessed 20/03/2014).

Cox, J. (2013) Observations on the Regulation of the Water Sector (Lecture, Royal Academy of Engineering, 5 March 2013) available at http://www.ofwat.gov.uk/mediacentre/speeches/prs_spe20130305jcrae.pdf (Accessed 27/03/2014).

DEA (2013) *Department of Environmental Affairs Annual Report 2012/13* available at https://www.environment.gov.za/sites/default/files/docs/environmental_affairs2012_2013_annualreport.pdf (Accessed 15/04/2014).

DEAT (undated) *Strategic Environmental Assessment*, Integrated Environmental Management Information Series available at https://www.environment.gov.za/sites/default/files/docs/series10_strategic_environmental_assessment.pdf (Accessed 20/03/2014).

DEFRA (2003) *Consultation Paper on Existing Private Sewers and Drains in England and Wales* Table 5.2 (archive document available from DEFRA on request).

DEFRA (2008) *A Strategy for Promoting an Integrated Approach to the Management of Coastal Areas in England* available at http://archive.defra.gov.uk/environment/marine/documents/protected/iczm/iczm-strategy-england.pdf (Accessed 19/03/2014).

DEFRA (2011) *Drought Permits and Drought Orders* including the Drought Direction 2011 available at https://www.gov.uk/government/uploads/system/uploads/attachment_data/file/69441/drought-permits-drought-orders.pdf (Accessed 27/03/2014).

DEFRA (2011a) *National Standards for Sustainable Drainage Systems* and related documents available at https://www.gov.uk/government/consultations/implementation-of-the-sustainable-drainage-provisions-in-schedule-3-to-the-flood-and-water-management-act-2010 (Accessed 27/03/2014).

DEFRA (2012) *The Water Resources Management Plan Direction 2012* available at http://www.defra.gov.uk/files/wrmp-direction2012.pdf (Accessed 19/03/2014).

DEFRA (2012a) *Company Social Tariffs: Guidance to Water and Sewerage Undertakers and the Water Services Regulation Authority under Section 44 of the Flood and Water Management Act 2010* available at https://www.gov.uk/government/uploads/system/uploads/attachment_data/file/69564/pb13787-social-tariffs-guidance.pdf (Accessed 27/03/2014).

DEFRA (2013) *Catchment Based Approach: Improving the Quality of our Water Environment* available at https://www.gov.uk/government/uploads/system/uploads/attachment_data/file/204231/pb13934-water-environment-catchment-based-approach.pdf (Accessed 19/03/2014).

DEFRA (2013a) *Making the Most of Every Drop: Consultation on Reforming the Water Abstraction Management System* available at https://consult.defra.gov.uk/water/abstraction-reform/ (Accessed 24/03/2014).

DEFRA/EA (2009) *Protecting our Water, Soil and Air: A Code of Good Agricultural Practice for Farmers, Growers and Land Managers* available at https://www.gov.uk/government/uploads/system/uploads/attachment_data/file/268691/pb13558-cogap-131223.pdf (Accessed 24/03/2104).

DEFRA/EA (2011) *The National Flood and Coastal Erosion Risk Management Strategy for England* available at http://www.official-documents.gov.uk/document/other/9780108510366/9780108510366.pdf (Accessed 20/03/2014).

DEFRA/OFWAT (2003) *Structure of the Water Industry in England: Does it Remain Fit for Purpose?* available at http://www.ofwat.gov.uk/publications/commissioned/rpt_com_pr04defraofwatexecsum.pdf (Accessed 27/03/2014).

DEFRA/RPA (2014) *The Guide to Cross Compliance in England 2014* available at http://rpa.defra.gov.uk/rpa/index.nsf/53f9502f3074384e80256cae00506fcc/6eb355ea8482ea61802573b1003d2469/$FILE/ATTUBLY9/The%20Guide%20to%20Cross%20Compliance%20in%20England%202014%20complete%20edition.pdf (Accessed 24/03/2014).

DEHP (2013) *Queensland Water Quality Guidelines 2009* available at http://www.ehp.qld.gov.au/water/pdf/water-quality-guidelines.pdf (Accessed 24/03/2014).

DEHP (2013a) *Annual Report on the Administration of the Environmental Protection Act 1994* available at http://www.ehp.qld.gov.au/about/corporatedocs/ (Accessed 10/04/2014).

DEHP (2014) *Queensland Environmental Offsets Policy* (not in effect) available at https://www.ehp.qld.gov.au/management/environmental-offsets/offsets-bill.html (Accessed 24/03/2014).

DEHP (undated) *Regulatory Strategy* available at http://www.ehp.qld.gov.au/management/planning-guidelines/policies/regulatory-strategy.html (Accessed 24/03/2014).

Department of Communities and Local Government (2010) *Code for Sustainable Homes* available at https://www.gov.uk/government/policies/improving-the-energy-efficiency-of-buildings-and-using-planning-to-protect-the-environment/supporting-pages/code-for-sustainable-homes (Accessed 27/03/2014).

Department of Housing and Public Works (2013) *Queensland Plumbing and Wastewater Code* available at http://www.hpw.qld.gov.au/SiteCollection Documents/QueenslandPlumbingAndWastewaterCode_15January2013.pdf (Accessed 27/03/2014).

Department of Housing and Public Works (2013a) *Queensland Development Code* available at http://www.hpw.qld.gov.au/construction/Building Plumbing/Building/BuildingLawsCodes/QueenslandDevelopmentCode/Pages/QueenslandDevelopmentCodeCurrentParts.aspx (Accessed 27/03/2014).

Department of Public Works (2000) *Guide for Architects Concerning Drainage, Water Supply and StormWater Drainage* available at http://www.publicworks.gov.za/PDFs/consultants_docs/DRAINAGESTORMWATER.PDF (Accessed 27/03/2014).

DERM (2010a) *Burnett Basin ROP 2003, amended April 2010* available at http://www.dnrm.qld.gov.au/__data/assets/pdf_file/0009/105102/burnett-rop.pdf (Accessed 21/03/2014).

DERM (2010b) *Total Water Cycle Management Planning Guideline* available at https://www.ehp.qld.gov.au/water/policy/pdf/twcmp-guideline.pdf (Accessed 26/03/2014).

DERM (2011) *Condamine and Balonne Resource Operations Plan Amendment Consultation Report* available at http://www.nrm.qld.gov.au/wrp/pdf/condamine/cb-rop-amend-consultdec11.pdf (Accessed 20/03/2014).

DERM (2012) *Water Resources in Queensland* available at http://www.nrm.qld.gov.au/water/water_in_qld.html (Accessed 07/04/2014).

DETR (1998) *The Review of Water Abstraction Licensing in England and Wales: A Consultation Paper*, archive document available from DEFRA on request.

DETR (1999) *Taking Water Responsibly* archive document available from DEFRA on request.

DEWS (2010a) *Planning Guidelines for Water and Sewerage* available at http://www.dews.qld.gov.au/water-supply-regulations/service-providers/forms-regulations (Accessed 27/03/2014).

DEWS (2010b) *Guidelines for Preparing Customer Service Standards* available at http://www.dews.qld.gov.au/water-supply-regulations/service-providers/forms-regulations (Accessed 27/03/2014).

DEWS (2012) *Queensland's Water Sector: A 30-Year Strategy: Discussion Paper: Shaping our Water Future* available at http://www.dews.qld.gov.au/policies-initiatives/water-sector-reform/30-year-strategy (Accessed 20/03/2014).

DEWS (2012a) *Bulk Water Supply Code* available at http://www.dews.qld.gov.au/__data/assets/pdf_file/0013/32305/bulk-water-supply-code.pdf (Accessed 27/03/2014).

DEWS (2013) *Water and Sewerage Services Code for Small Customers in South East Queensland* available at http://www.dews.qld.gov.au/__data/assets/pdf_file/0014/32306/water-sewerage-services-code.pdf (Accessed 27/03/2014).

DEWS (2013a) *Queensland Urban Drainage Manual* 3rd Edition available at http://www.dews.qld.gov.au/water-supply-regulations/urban-drainage (Accessed 27/03/2014).

DHS (2012) *Ministerial Sanitation Task Team Report: July 2012 Review, Investigation and Evaluation of the National Sanitation Programme – Towards Continuous Improvement* available at http://www.dhs.gov.za/sites/default/files/documents/publications/Sanitation_Task_Team_Report_July2012.pdf (Accessed 26/03/2014).

DHS (2013) *Annual Report 2012/13* available at http://www.dhs.gov.za/content.php?pagename=Annual-Reports (Accessed 27/03/2014).

DNRM (2012) *Water Management Plan for the Upper Condamine Alluvium Sustainable Diversion Limit Area: Interim Water Resource Plan for the Commonwealth Water Act 2007* available at http://www.dnrm.qld.gov.au/__data/assets/pdf_file/0019/104842/iwrp-uca-sdl-watermgmtplan.pdf (Accessed 20/03/2014).

DNRM (2013) *Annual Report 2012–2013 Queensland's Water Resource Plans: Summary* (full documents on request) available at http://www.dnrm.qld.gov.au/water/catchments-planning/planning-process/annual-report-wrp (Accessed 20/03/2014).

DNRM (undated) *Queensland Regional Natural Resource Management Investment Programme 2013–2018* available, with supporting information, at http://www.regionalnrm.qld.gov.au/ (Accessed 20/03/2014).

DPMAG (undated) *Rural Diffuse Pollution Plan for Scotland* available at http://www.sepa.org.uk/water/diffuse_pollution.aspx (Accessed 26/03/2014).

Driver, A. *et al.* (2011) *Implementation Manual for Freshwater Ecosystem Priority Areas*. WRC Report 1801/1/11.

Dublin Statement (1992) *Dublin Statement on Water and Sustainable Development* available at http://www.wmo.ch/pages/prog/hwrp/documents/english/icwedece.html (Accessed 17/03/2014).

DWA (2003a) *Sanitation is Dignity: Protocol to Manage the Potential of Groundwater Contamination from Onsite Sanitation* available at http://www.dwaf.gov.za/dir_ws/wspd/policylist.aspx (Accessed 27/03/2014).

DWA (2010) *Groundwater Strategy* available at http://www.dwaf.gov.za/Groundwater/gs.aspx (Accessed 19/03/2014).

DWA (2010a) *Water Services Regulation Strategy* available on request from DWA and on file with the author.

DWA (2012) *Green Drop Report* available at http://www.dwa.gov.za/Dir_WS/GDS/Docs/DocsDefault.aspx (Accessed 27/03/2014).

DWA (2013) *National Water Resource Strategy* 2nd Edition available at http://www.dwaf.gov.za/nwrs/ (Accessed 20/03/2014).

DWA (2013a) *Revised Water Pricing Strategy for Raw Water III Draft* available at http://www.dwaf.gov.za/Projects/PERR/documents/Revised%20Pricing%20Strategy%2027-09-2013.pdf (Accessed 24/03/2014).

DWA (2013b) Media statement 'The Progress Thus Far with Regard to a Response to the Matter of Acid Mine Drainage Particularly in the Witwatersrand Area 26 November 2013' available at http://www.dwaf.gov.za/Communications/PressReleases/2013/MEDIA%20RELEASE%20ON%20AMD%20PROGRESS%20261113.pdf (Accessed 24/03/2014).

DWA (2013c) *Annual Report 2012/13* available at http://www.dwaf.gov.za/media.aspx (Accessed 09/04/2014).

DWA (2013d) Media statement 'Government Working Proactively on Water Issues' available at http://www.dwaf.gov.za/Communications/PressReleases/2014/Government%20proactively%20working%20on%20water%20issues.pdf (Accessed 26/03/2014).

DWA (undated) *The Blue Drop and No Drop Handbook* available at http://www.dwaf.gov.za/dir_ws/DWQR/subscr/ViewComDoc.asp?Docid=604 (Accessed 27/03/2014).

DWAF (1994) *Water Supply and Sanitation Policy* available at http://www.dwaf.gov.za/Documents/ (Accessed 6/4/2007).

DWAF (1996) *South African Water Quality Guidelines* available at http://www.dwaf.gov.za/iwqs/wq_guide/index.asp (Accessed 24/03/2014).

DWAF (1997) *National Water Policy for South Africa White Paper* available at http://www.dwaf.gov.za/Documents/default.aspx?type=policy (Accessed 20/03/2014).

DWAF (1999) *Water Resource Protection Policy Implementation: Resource Directed Measures for Protection of Water Resources* available at http://www.dwaf.gov.za/Documents/Policies/WRPP/default.htm (Accessed 19/03/2014).

DWAF (1999a) *Water Resource Protection Policy Volume 6: Groundwater* available at http://www.dwaf.gov.za/Documents/Policies/WRPP/default.htm (Accessed 24/03/2014).

DWAF (2001) *White Paper on Basic Household Sanitation* available at www.dwaf.gov.za/Documents/Policies/SanitationReviewPolicy.pdf (Accessed 27/03/2014).

DWAF (2003) *Strategic Framework for Water Services* available at http://www.dwaf.gov.za/Documents/ (Last accessed 6/4/2007).

DWAF (2004) *National Water Resources Strategy* (NWRS) available at http://www.dwaf.gov.za/Documents/default.aspx?type=policy (Accessed 20/03/2014).

DWAF (2004a) *Policy on Financial Assistance to Resource Poor Irrigation Farmers* available at http://www.dwaf.gov.za/Documents/Policies/rpf/RPFAssistanceSep04.pdf (Accessed 19/03/2014).

DWAF (2004b) *Guidelines for Stakeholder Participation in IWRM* available at http://www.iwrm.co.za/resource%20doc/irwm%201/Stakeholder%20Participation/Guidelines/GUIDELINES%20FOR%20STAKEHOLDER%20PARTICIPATION%20LEVEL%203.pdf (Accessed 20/03/2014).

DWAF (2004c) *National Water Conservation and Water Demand Management Strategy* available at http://www.dwaf.gov.za/docs/Other/WaterUse Conservation/WCWDMAug04.pdf (Accessed 27/03/2014).

DWAF (2004d) *Water Conservation and Water Demand Management Strategy for the Water Services Sector* available at http://www.dwaf.gov.za/

docs/Other/WaterUseConservation/WCWDMWaterServicesAug04.pdf (Accessed 27/03/2014).

DWAF (2005) *Model Bylaws Pack. Model Credit Control and Debt Collection Bylaws; Model Water Services Bylaws* available at http://intertest.dwaf.gov.za/dir_ws/waterpolicy/toolbox/policy_detail_print.asp?Policy=422 (Accessed 27/03/2014).

DWAF (2005a) *A Drinking Water Quality Framework for South Africa* available at http://www.dwaf.gov.za/Documents/Other/DWQM/DWQMFrameworkDec05.pdf (Accessed 27/03/2014).

DWAF (2005b) *Drinking Water Quality Regulation Strategy* available at http://www.dwaf.gov.za/events/WWMD/2007/documents/DWQReg-Sep05.pdf (Accessed 27/03/2014).

DWAF (2006) *Guidelines for the Utilisation and Disposal of Wastewater Sludge* available at http://www.dwaf.gov.za/Dir_WQM/docs/SewageSludgeMar06vol1.asp (Accessed 27/03/2014).

DWAF (2007) *Resource Directed Management of Water Quality Training Manual* available at http://www.dwaf.gov.za/Dir_WQM/RDMDocs/RDWQO%20RWQO%20training%20manual.pdf (Accessed 19/03/2014).

DWAF (2007a) *Guidelines for Catchment Management Strategies* available at http://www.dwaf.gov.za/Documents/Other/CMA/CMSFeb07/CMSFeb07Ed1.pdf (Accessed 20/03/2014).

DWAF (2007b) *Technical Guidelines for Development of the Water Resource Classification System (WRCS) Volume 1* available at http://www.dwaf.gov.za/RDM/documents/vol01complete.pdf (Accessed 24/03/2014).

DWAF (2007c) *Free Basic Water: Implementation Strategy 2007 Consolidating and Maintaining* available at http://www.dwaf.gov.za/dir_ws/wspd/policylist.aspx.

DWAF (2008) *Strategy for Water Allocation Reform* available at http://www.dwaf.gov.za/WAR/rules.aspx (Accessed 27/03/2014).

DWAF (2008a) *A Guideline for the Assessment, Planning and Management of Groundwater Resources in South Africa* available at http://www.dwaf.gov.za/Groundwater/Documents.aspx (Accessed 24/03/2014).

DWAF (2008b) *Free Basic Sanitation Implementation Strategy* available at http://www.dwaf.gov.za/dir_ws/wspd/policyinfo.aspx?filen=556 (Accessed 27/03/2014).

DWAF (Undated) *Establishing a Catchment Management Agency* available at http://www.dwaf.gov.za/CM/cma.htm (Accessed 20/03/2014).

DWQR (2012) *Drinking Water Quality in Scotland 2012* available at http://www.dwqr.org.uk/information/annual-report.htm (Accessed 27/03/2014).

EA (2005) *Water for Life and Livelihoods: Strategy for River Basin Planning: Consultation* archive document available from the EA on request.

EA (2006) *Do We Need Large Scale Water Transfers for South East England?* archive document available from the EA on request.

EA (2006a) *Water for Life and Livelihoods: A Framework for River Basin Planning in England and Wales* available at http://cdn.environment-agency.gov.uk/geho0506bkvx-e-e.pdf (Accessed 19/03/2014).

EA (2009) *Water for People and the Environment: A National Water Resources Strategy for England and Wales* available at http://www.environment-agency.gov.uk/research/library/publications/40731.aspx (Accessed 19/03/2014).

EA (2009a) *Water for People and the Environment: Water Resources Strategy Regional Action Plan for Anglian Region* available at http://www.environment-agency.gov.uk/research/library/publications/40731.aspx (Accessed 20/03/2014).

EA (2009b) *River Basin Management Plans* available at http://www.environment-agency.gov.uk/research/planning/148254.aspx (Accessed 19/03/2014).

EA (2010) *Water for People and the Environment: Water Resources Action Plan for England and Wales* available at http://www.environment-agency.gov.uk/research/library/publications/40731.aspx (Accessed 19/03/2014).

EA (2010a) *SR10 No.3 Discharge to Surface Water: Secondary Treated Domestic Sewage with a Maximum Daily Volume between 5 and 20 Cubic Metres per Day* available at http://www.environment-agency.gov.uk/business/topics/permitting/117722.aspx (Accessed 26/03/2014).

EA (2011) *The Case for Change* available at http://cdn.environment-agency.gov.uk/geho1111bvep-e-e.pdf (Accessed 03/04/2014).

EA (2011a) *Enforcement and Sanctions: Guidance* available at http://www.environment-agency.gov.uk/business/regulation/31851.aspx (Accessed 24/03/2014).

EA (2011b) *Water Company Drought Plan Guideline* available at http://a0768b4a8a31e106d8b0-50dc802554eb38a24458b98ff72d550b.r19.cf3.rackcdn.com/geho0311btoj-e-e.pdf (Accessed 27/03/2014).

EA (2012) *River Basin Management: Working Together England and Wales* available at http://www.environment-agency.gov.uk/research/planning/33248.aspx (Accessed 19/03/2014).

EA (2012a) *Environment Agency Abstraction Charges Scheme 2013/14* available at http://www.environment-agency.gov.uk/business/regulation/38809.aspx (Accessed 24/03/2014).

EA (2012b) *Groundwater Protection Principles and Practice* available at http://www.environment-agency.gov.uk/research/library/publications/144346.aspx (Accessed 26/03/2014).

EA (2013) *Managing Water Abstraction* available at http://www.environment-agency.gov.uk/business/topics/water/119927.aspx (Accessed 19/03/2014).

EA (2013a) *National Environment Programme* available from the Agency on request http://www.environment-agency.gov.uk/business/sectors/33065.aspx (Accessed 19/03/2014).

EA (2013b) *Water for Life and Livelihoods* available at http://www.environment-agency.gov.uk/research/planning/149119.aspx (Accessed 19/03/2014).

EA (2013c) *England's Waters: Challenges and Choices: Summary of Significant Water Management Issues: A Consultation* available with regional consultations at http://www.environment-agency.gov.uk/research/planning/33252.aspx (Accessed 19/03/2014).

EA (2013d) *Water Stressed Areas Final Classification* available at https://www.gov.uk/government/uploads/system/uploads/attachment_data/file/244333/water-stressed-classification-2013.pdf (Accessed 27/03/2014).

EA (2014) *Request for information under Freedom of Information Act/ Environmental Information Regulations NRMAR110* (on file with the author).

EA (undated) *Changing Water Abstraction & Impoundment Licences* available at http://www.environment-agency.gov.uk/business/topics/water/135345.aspx (Accessed 24/03.2014).

EA/OFWAT/DEFRA/Welsh Government (2012) *Water Resources Planning Guideline* available at http://cdn.environment-agency.gov.uk/geho0612bwpe-e-e.pdf (Accessed 19/03/2014).

EA/Scottish Government (2009) *The River Basin Management Plan for the Solway Tweed River Basin District 2009–2015* available at http://www.sepa.org.uk/water/river_basin_planning.aspx (Accessed 19/03/2014).

Earle, A. *et al.* (2005) *Domestic Water Provision in the Democratic South Africa* available at http://www.acwr.co.za/pdf_files/02.pdf (Accessed 03/04/2014).

European Commission (1993) *Towards Sustainability: EC Fifth Environmental Action Programme* OJ 1993 C138/1.

European Commission (2002) *Guidance on Public Participation in Relation to the Water Framework Directive: Active Involvement, Consultation and Public Access to Information* available at http://ec.europa.eu/environment/water/water-framework/facts_figures/guidance_docs_en.htm (Accessed 19/03/2014).

European Commission (2006) *Communication on a Thematic Strategy for Soil Protection* COM (2006) 231 final.

European Commission (2006a) *Proposal for a Directive Establishing a Framework for Protection of Soil* COM (2006) 232 final available at http://ec.europa.eu/environment/soil/index.htm (Accessed 27/03/2014).

European Commission (2012) *COMMISSION STAFF WORKING DOCUMENT European Overview (2/2) Accompanying the document REPORT FROM THE COMMISSION TO THE EUROPEAN PARLIAMENT AND THE COUNCIL on the Implementation of the Water Framework Directive (2000/60/EC) River Basin Management Plans* SWD (2012) 379 final available at http://ec.europa.eu/environment/water/blueprint/ (Accessed 19/03/2014).

European Commission (2012a) *COMMISSION STAFF WORKING DOCUMENT European Overview (1/2) Accompanying the document REPORT FROM THE COMMISSION TO THE EUROPEAN PARLIAMENT AND THE COUNCIL on the Implementation of the Water Framework Directive (2000/60/EC) River Basin Management Plans* SWD (2012) 379 final available at http://ec.europa.eu/environment/water/blueprint/ (Accessed 19/03/2014).

European Commission (2012b) *River Basin Management Plans: REPORT on the Implementation of the Water Framework Directive (2000/60/EC)* COM 2012 670 available at http://ec.europa.eu/environment/water/blueprint/ (Accessed 19/03/2014).

European Commission (2012c) *A Blueprint to Safeguard Europe's Water Resources* COM (2012) 673 available at http://ec.europa.eu/environment/water/blueprint/ (Accessed 19/03/2014).

European Commission (2012d) *COMMISSION STAFF WORKING DOCU-MENT Accompanying the document REPORT FROM THE COMMISSION TO THE EUROPEAN PARLIAMENT AND THE COUNCIL on the Implementation of the Water Framework Directive 2006/60/EC River Basin Management Plans SWD (2012) 379 final, Member State: United Kingdom 29/30* available at http://ec.europa.eu/environment/water/blueprint/ (Accessed 19/03/2014).

European Commission (2012e) *COMMISSION STAFF WORKING DOCU-MENT IMPACT ASSESSMENT Accompanying the document COMMUNI-CATION FROM THE COMMISSION TO THE EUROPEAN PARLIAMENT, THE COUNCIL, THE EUROPEAN ECONOMIC AND SOCIAL COMMITTEE AND THE COMMITTEE OF THE REGIONS A Blueprint to Safeguard Europe's Water Resources SWD (2012) 382 final* available at http://ec.europa.eu/environment/water/blueprint/ (Accessed 19/03/2014).

European Commission (2013) *Proposal for a Directive Establishing a Framework for Maritime Spatial Planning and Integrated Coastal Management* COM (2013) 133 final.

European Commission (2013a) *Seventh Report on the Implementation of the Urban Waste Water Treatment Directive (91/271/EEC)* COM (2013) 574.

Explanatory Notes to the Land, Water and Other Legislation Amendment (Qld) Bill (2013) available at https://www.legislation.qld.gov.au/Bills/54PDF/2013/LandWaterOLAB13E.pdf (Accessed 19/03/2014).

FAO (1999) *Issues in Water Law Reform* Legislative Series 67 FAO Rome.

Forestry Commission (2011) *Forests and Water* available at http://www.forestry.gov.uk/website/forestry.nsf/byunique/infd-8bvgx9 (Accessed 24/03/2014).

Fortson, D. (2014) How Water Barons Siphoned off Profits. *Sunday Times*, 26 January.

Government of South Africa (2013) *YearBook 2012/13 (Water)* available at http://www.gcis.gov.za/sites/www.gcis.gov.za/files/docs/resourcecentre/yearbook/2012/24%20Water%20Affairs.pdf.

Grafton, R. Q. *et al.* (2010a) *An Integrated Assessment of Water Markets: Australia, Chile, China, South Africa and the USA* NBER Working Paper 16203.

Grafton, R. Q. *et al.* (2010) *Water Markets: Australia's Murray–Darling Basin and the UU Southwest* NBER Working Paper 15797.

GWP/INBO (2009) *A Handbook for Integrated Water Resources Management in Basins* available at http://www.unwater.org/downloads/GWP-INBOHandbookForIWRMinBasins.pdf (Accessed 19/03/2014).

Hansard House of Commons [Vol.413] Col.32.

Hansard House of Lords [Vol.654] Col.972.

Hansard House of Lords [Vol.752] Col.106–109.

HarmoniCOP (2005) *Learning Together to Manage Together: Improving Participation in Water Management* available at http://www.harmonicop.uni-osnabrueck.de/HarmoniCOPHandbook.pdf (Accessed 19/03/2014).

Helmer, R. and Hespanhol, I. (Eds.) (1997) *Water Pollution Control: A Guide to the Use of Water Quality Management Principles*. UNEP/WHO available at http://www.who.int/water_sanitation_health/resources/resquality/en/index.html (Accessed 25/03/2014).

HM Government (2012) *Water for Life* Cmnd.8230.

HM Government (2013) *Government Response to the EFRA Committee's Pre-Legislative Scrutiny of the Draft Water Bill* Cmnd.8643.

HM Government Building Regulations (2010) Parts G and H available at http://www.planningportal.gov.uk/buildingregulations/approveddocuments/downloads (Accessed 27/03/2014).

House of Representatives Standing Committee on Regional Australia (2011) *Of Drought and Flooding Rains Inquiry into the Impact of the Guide to the Murray–Darling Basin Plan* available at http://www.aph.gov.au/Parliamentary_Business/Committees/House_of_Representatives_Committees?url=ra/murraydarling/report.htm (Accessed 20/03/2014).

Inkomati CMA (undated) *The Inkomati Catchment Management Strategy* and associated documents available at http://inkomaticma.co.za/publications/icma-documents.html# (Accessed 19/03/2014).

Kessides, I. (2004) *Reforming Infrastructure: Privatisation, Regulation and Competition* WB Policy Research Report WB/OUP available at http://documents.worldbank.org/curated/en/2004/01/4297210/reforming-infrastructure-privatization-regulation-competition (Accessed 27/03/2014).

Kleynhans, C. and Louw, M. D. (2007) EcoClassification and EcoStatus Determination in *River EcoClassification: Manual for EcoStatus Determination (version 2)* WRC/DWAF available at http://www.dwaf.gov.za/Documents/Policies/WRPP/default.htm (Accessed 24/03/2014).

Krinner, W. *et al.* (1999) *Sustainable Water Use in Europe: Part 1 Sectoral Water Use* EEA Copenhagen available at http://reports.eea.europa.eu/binaryeenviasses01pdf/en (Accessed 03/04/2014).

Macrory, R. (2002) *Modernising Environmental Justice: Regulation and the Role of an Environmental Tribunal* UCL available at http://www.ucl.ac.uk/laws/environment/tribunals/docs/full_report.pdf.

Marin, P. (2009) *Public-Private Partnerships for Urban Water Utilities: A Review of Experiences in Developing Countries* World Bank.

Miranda, L. R. (2014) 'South Africans Protest at Lack of Basic Services' The Real Agenda News, 11 February available at http://real-agenda.com/2014/02/11/south-africans-protest-for-lack-of-basic-services/ (Accessed 26/03/2014).

Mjoli, N. (2009) *Review of Sanitation Policy and Practice in South Africa 2001–2008* WRC 1741/1/09 available at http://www.wrc.org.za/Knowledge%20Hub%20Documents/Research%20Reports/1741-1-09%20Developing%20Communities.pdf (Accessed 27/03/2014).

NHMRC/NRMMC (2011) (as updated, rolling revision programme) *Australian Drinking Water Guidelines* available at http://www.nhmrc.gov.au/publications/synopses/eh19syn.htm (Accessed 27/03/2014).

NWC (2011) *The National Water Initiative: Securing Australia's Water Future 2011 Assessment* available at http://www.nwc.gov.au/nwi/biennial-assessments (Accessed 24/03/2014).

NWC (2012) *Australian Water Markets Report 2011/12* available at http://nwc.gov.au/home/publications/water-markets-11-12 (Accessed 24/03/2014).

OECD (2013) *Environmental Performance Reviews: South Africa 2013* OECD Publishing.

OFWAT (2012) *Delivering Proportionate and Targeted Regulation: Ofwat's Risk-Based Approach* available at http://www.ofwat.gov.uk/regulating/compliance/pap_pos1203regcomp.pdf (Accessed 27/03/2014).

Pitt, M. (2008) *Learning the Lessons* available at http://webarchive.nationalarchives.gov.uk/20100807034701/http:/archive.cabinetoffice.gov.uk/pittreview/thepittreview/final_report.html (Accessed 20/03/2014).

Principles of Charging for Water Services 2010–2015. Statement available at http://www.watercommission.co.uk/view_Ministerial_Guidance.aspx (Accessed 27/03/2014).

Productivity Commission (2003) *Water Rights in Australia and Overseas* available at http://www.pc.gov.au/research/commission/water-rights (Accessed 16/04/2014).

Productivity Commission (2010) *Market Mechanisms for Recovering Water in the Murray–Darling Basin* available at http://www.pc.gov.au/__data/assets/pdf_file/0019/96004/water-recovery-report.pdf (Accessed 16/04/2014).

Pruss-Ustun, A. *et al.* (2008) *Safer Water Better Health* WHO available at http://whqlibdoc.who.int/publications/2008/9789241596435_eng.pdf (Accessed 17/03/2014).

Public Private Infrastructure Advisory Facility and Water and Sanitation Programme (2001) *New Designs for Water and Sanitation Transactions: Making Water Work for the Poor* available at https://www.ppiaf.org/sites/ppiaf.org/files/publication/WSP%20-%20New%20Designs%20Water%20Sanitation%20Transaction%202002.pdf (Accessed 27/03/2014).

QCA (2000) *Statement of Regulatory Pricing Principles for the Water Sector* available at http://www.qca.org.au/Water/Queensland-wide-issues/Water-Pricing-Principles/Final-Report/Water-Pricing-Principles#finalpos (Accessed 27/03/2014).

QCA (2013) *Position Paper: Long Term Regulatory Framework for SEQ Water Entities* and related documentation available at http://www.qca.org.au/Water/Urban-retail-water/Retail/SEQ-Reg-framework/In-Progress/SEQ-Reg-framework-principles (Accessed 27/03/2014).

Queensland Floods Commission of Inquiry 2012 Final Report available at http://www.floodcommission.qld.gov.au/__data/assets/pdf_file/0007/11698/QFCI-Final-Report-March-2012.pdf (Accessed 19/03/2014).

Queensland Government (2006) *Central Queensland Regional Water Supply Strategy* available at http://www.dews.qld.gov.au/water-supply-regulations/regional-water-supply (Accessed 27/03/2014).

Queensland Government (2009) *South East Queensland Natural Resource Management Plan* available at http://www.seqcatchments.com.au/resources-nrm-plan.html.

Queensland Government (2010) *Far North Queensland Regional Water Supply Strategy* available at http://www.dews.qld.gov.au/water-supply-regulations/regional-water-supply (Accessed 27/03/2014).

Queensland Government (2012) *North West Queensland Draft Regional Water Supply Strategy* available at http://www.dews.qld.gov.au/water-supply-regulations/regional-water-supply (Accessed 27/03/2014).

Queensland Water Commission (2010) *South East Queensland Water Strategy* available at http://www.dews.qld.gov.au/water-supply-regulations/regional-water-supply (Accessed 27/03/2014).

Rieu-Clarke, A. and Allan, A. (2008) *Role of Water Law: Assessing Governance in the Context of IWRM* Report No.D6.3 EU 'STRIVER' Project available at http://kvina.niva.no/striver/Disseminationofresults/STRIVER Reports/tabid/80/Default.aspx (Accessed 19/03/2014).

River Basin Districts Typology, Standards and Groundwater Threshold Values (Water Framework Directive) (England and Wales) Directions 2009 available at http://archive.defra.gov.uk/environment/quality/water/wfd/documents/river-basin-typology-standards.pdf (Accessed 26/03/2014).

Rogers, P. and Hall, A. (2000) *Integrated Water Resources Management* GWP TAC Background Paper No.4 available at http://www.gwp.org/en/The-Challenge/IWRM-Resources/ (Accessed 19/03/2014).

Rogers, P. and Hall, A. (2013) *Effective Water Governance* GWP TAC Background Paper No. 7 available at http://www.gwp.org/en/The-Challenge/IWRM-Resources/ (Accessed 19/03/2014).

Roper, H. *et al.* (2006) *Stranded Irrigation Assets* Australian Government Productivity Commission Staff Working Paper available at http://www.pc.gov.au/research/staff-working/strandedirrigation (Accessed 24/03/2014).

SANS 241: Specifications for Drinking Water available at https://www.sabs.co.za/ (Accessed 10/04/2014).

Scotland River Basin District (Surface Water Typology, Environmental Standards, Condition Limits and Groundwater Threshold Values) Directions 2009 available at http://www.scotland.gov.uk/Resource/Doc/298071/0092869.pdf (Accessed 24/03/2014).

Scottish Executive (2001) *The Water Services Bill: The Executive's Proposals* available at http://www.scotland.gov.uk/Publications/2001/03/8747/File-1 (Accessed 27/03/2014).

Scottish Executive (2002) *The Future for Scotland's Waters: Proposals for Legislation* available at http://www.scotland.gov.uk/Publications/2003/04/17036/21508 (Accessed 24/03/2014).

Scottish Executive (2004) *A Strategy for Scotland's Coast and Inshore Waters* available at http://www.scotland.gov.uk/Publications/2004/07/19639/40165 (Accessed 19/03/2014).

Scottish Executive (2004a) *Investing in Water Services 2006–2014* available at http://www.scotland.gov.uk/Publications/2004/07/19648/40230 (Accessed 27/03/2014).

Scottish Executive (2004b) *Paying for Water Services 2006–2010* available at http://www.scotland.gov.uk/Publications/2004/07/19649/40255 (Accessed 27/03/2014).

Scottish Executive (2005) *Investing in Water Services: Objectives for 2006–2014 Statement by the Scottish Executive* available at http://www.watercommission.co.uk/UserFiles/Documents/Statement%20on%20investment%20objectives.pdf (Accessed 27/03/2014).

Scottish Executive (2005a) *Prevention of Environmental Pollution from Agricultural Activity: A Code of Good Practice* available at http://www.sepa.org.uk/land/agriculture/agricultural_guidance.aspx (Accessed 24/03/2014).

Scottish Executive (2005b) *The Water Environment (Controlled Activities) (Scotland) Regulations 2005: Policy Statement and Regulatory Impact Assessment* Paper 2005/10 available at http://www.scotland.gov.uk/Publications/2005/05/0995747/57481 (Accessed 24/03/2014).

Scottish Executive (2006) *Implementing the Water Environment and Water Services (Scotland) Act 2003: The Designation of Responsible Authorities: A Policy Statement* available at http://www.scotland.gov.uk/Publications/2006/03/08143009/0 (Accessed 19/03/2014).

Scottish Executive (2006a) *Implementing the Water Environment and Water Services (Scotland) Act 2003: Process for Taking Into Consideration Third Party Representations in Connection with Applications under the Water Environment (Controlled Activities) (Scotland) Regulations 2005* available at http://www.scotland.gov.uk/Publications/2006/12/13144359/1 (Accessed 24/03/2014).

Scottish Executive (2006b) *Planning Advice Note 79 Water and Drainage* available at http://www.scotland.gov.uk/Publications/2006/09/26152857/15 (Accessed 27/03/2014).

Scottish Government (2009) *The River Basin Management Plan for the Scotland River Basin District 2009–2015* available at http://www.sepa.org.uk/water/river_basin_planning.aspx (Accessed 19/03/2014).

Scottish Government (2010) *Building a Hydro-Nation Consultation* available at http://www.scotland.gov.uk/Publications/2010/12/14111932/0 (Accessed 27/03/2014).

Scottish Government (2012) *Scotland the Hydro Nation: Prospectus and Proposals for Legislation* available at http://www.scotland.gov.uk/Publications/2012/02/9536 (Accessed 27/03/2014).

Scottish Government (2012a) *Investing in and Paying for Your Water Services from 2015* available at http://www.scotland.gov.uk/Resource/0039/00395226.pdf (Accessed 27/03/2014).

Scottish Government (2013) *Implementing the Water Environment and Water Services (Scotland) Act 2003: Environmental Standards for the Water Environment: A Consultation* available at http://www.scotland.gov.uk/Resource/0043/00439748.pdf (Accessed 26/03/2014).

Scottish Parliament Finance Committee (2002) *Report on the Financial Memorandum for the Water Environment and Water Services Bill* available at http://www.scottish.parliament.uk/parliamentarybusiness/PreviousCommittees/15877.aspx (Accessed 20/03/2014).

Scottish Parliament Transport and Environment Committee (2002) *16th Report: Stage 1 Report Water Environment and Water Services Bill* available at http://archive.scottish.parliament.uk/business/committees/historic/x-transport/2002.htm (Accessed 19/03/2014).

Scottish Water (2009) *Water Resource Plan* available at http://www.scottish-water.co.uk/assets/about%20us/files/key%20publications/adoptedwrp09summarydoc.pdf (Accessed 19/03/2014).

Scottish Water (2009a) *Sustainable Land Management* available at http://www.scottishwater.co.uk/assets/about%20us/files/corporate%20responsibility/swincentiveschemeinfo.pdf (Accessed 26/03/2014).

Scottish Water (2012) *Code of Practice* available at http://www.scottishwater.co.uk/contact-us/our-promise-to-you/customer-charter (Accessed 27/03/2014).

Scottish Water (2013) *Strategic Projection* available at http://www.scottishwater.co.uk/About-Us/Publications/Strategic-Projections (Accessed 27/03/2014).

Scottish Water (2013a) *Draft Business Plan 2015 – 2021* available at http://www.scottishwater.co.uk/About-Us/Publications/Strategic-Projections (Accessed 27/03/2014).

Scottish Water (Objectives for 2010 – 2015) Directions 2009 available at http://www.watercommission.co.uk/view_Ministerial_Guidance.aspx (Accessed 27/03/2014).

SEPA (2009) *The River Basin Management Plan for the Scotland River Basin District: Summary of Responses to the Draft Plan Consultation* available at http://www.sepa.org.uk/water/river_basin_planning/early_basin_planning_work.aspx (Accessed 19/03/2014).

SEPA (2009a) *The River Basin Management Plan for the Solway Tweed River Basin District: Summary of Responses to the Draft Plan Consultation* available at http://www.sepa.org.uk/water/river_basin_planning/early_basin_planning_work.aspx (Accessed 19/03/2014).

SEPA (2009b) *Groundwater Protection Policy 19 V3* available at http://www.sepa.org.uk/water/water_regulation/regimes/groundwater/planning.aspx (Accessed 24/03/2014).

SEPA (2012) *Working Together to Protect and Improve Scotland's Water Environment: Getting Involved in Developing the Second River Basin Management Plan Consultation* available at http://www.sepa.org.uk/about_us/consultations/closed_consultations.aspx (Accessed 19/03/2014).

SEPA (2012a) *SEPA Enforcement Report 2011–2012* available at http://www.sepa.org.uk/about_us/publications/enforcement_reports.aspx (Accessed 24/03/2014).

SEPA (2013) *River Basin Planning: Developing the Second River Basin Management Plans* available at http://www.sepa.org.uk/water/river_basin_planning.aspx (Accessed 19/03/2014).

SEPA (2013a) *Current Condition and Challenges for the Future* available at http://www.sepa.org.uk/water/river_basin_planning/significant_issues.aspx (Accessed 19/03/2014).

SEPA (2013b) *CAR: A Practical Guide* available at http://www.sepa.org.uk/water/water_regulation.aspx (Accessed 24/03/2014).

SEPA (2013c) *Water Environment (Controlled Activities) Fees and Charges (Scotland) Scheme 2013* available at http://www.sepa.org.uk/about_us/charging_schemes/current_charging_schemes.aspx (Accessed 24/03/2014).

Sky News (2014, 13 February) 'South Africa Water Protesters "Shot By Police"' (online) 13 February, available at http://news.sky.com/story/1194450/south-africa-water-protesters-shot-by-police (Accessed 26/03/2014).

Solanes, M. and Jouravlev, A. (2007) *Revisiting Privatisation, Foreign Investment, Arbitration, and Water* available at http://www.eclac.cl/publicaciones/xml/0/32120/lcl2827e.pdf (Accessed 26/03/2014).

Solway Tweed River Basin District (Surface Water Typology, Environmental Standards, Condition Limits and Groundwater Threshold Values) Directions 2009 available at http://www.scotland.gov.uk/Resource/Doc/296383/0092090.pdf (Accessed 24/03/2014).

Statistics South Africa (2011) *Census 2011: Census in Brief* available at https://www.statssa.gov.za/Census2011/Products.asp (Accessed 27/03/2014).

Still, D. *et al.* (2007) *The Status and Use of Potable Water Efficient Devices in the Domestic and Commercial Environments in South Africa* WRC K5/1606 available at http://www.waterefficiencysa.co.za/wed%20reports/WEDReportExSummplusChapte1.pdf (Accessed 27/03/2014).

Taylor, L. (2010) Farmers Fear Water Act Priority *Sydney Morning Herald*, 16 October, available at http://www.smh.com.au/environment/water-issues/farmers-fear-water-act-priority-20101015-16nmo.html (Accessed 20/03/2014).

Tissington, K. (2011) *Basic Sanitation in South Africa: A Guide to Legislation, Policy and Practice* SERI available at http://www.seri-sa.org/index.php/research-7/resource-guides (Accessed 27/03/2014).

Transparency International (2008) *Global Corruption Report 2008* available at http://archive.transparency.org/publications/gcr/gcr_2008 (Accessed 26/03/2014).

Turner, G./CentreForum (2013) *Money Down the Drain: Getting a Better Deal for Customers in the Water Industry* available at http://www.centreforum.org/assets/pubs/money-down-the-drain.pdf (Accessed 27/03/2014).

UNDESA (2012) *World Population Prospects 2012 Revision* available at http://esa.un.org/unpd/wpp/Documentation/pdf/WPP2012_Volume-I_Comprehensive-Tables.pdf (Accessed 17/03/2014).

UNDP (1997) *Governance for Sustainable Human Development* available at http://mirror.undp.org/magnet/policy/ (Accessed 19/03/2014).

UNDP (2006) *Human Development Report: Beyond Scarcity: Power, Poverty and the Global Water Crisis* available at http://hdr.undp.org/sites/default/files/reports/267/hdr06-complete.pdf (Accessed 17/03/2014).

UNECE (2000) *Draft Guidelines on Sustainable Flood Prevention ECOSOC MP.WAT/2000/7* available at http://www.unece.org/fileadmin/DAM/env/water/publications/documents/guidelinesfloode.pdf (Accessed 19/03/2014).

UNEP (2008) *Vital Water Graphics* available at http://www.unep.org/dewa/vitalwater/index.html (Accessed 17/03/2014).

UNESCO (2009) *IWRM Guidelines at River Basin Level* available at http://www.gwptoolbox.org/ (Accessed 19/03/2014).

United Utilities (2013) *Draft Water Resources Management Plan 2013* available at http://corporate.unitedutilities.com/Water-Resources-Management-Plan.aspx (Accessed 19/03/2014).

UN-Water (2003) *World Water Development Report 1: Water for People, Water for Life* available at http://www.unesco.org/water/wwap/wwdr/ (Accessed 17/03/2014).

UN-Water (2006) *World Water Development Report 2: Water, a Shared Responsibility* available at http://www.unesco.org/water/wwap/wwdr/ (Accessed 17/03/2014).

UN-Water (2009) *World Water Development Report 3: Water in a Changing World* available at http://www.unesco.org/water/wwap/wwdr/ (Accessed 17/03/2014).

UN-Water (2012) *Status Report on The Application of Integrated Approaches to Water Resources Management* available at http://www.unwater.org/downloads/UNW_status_report_Rio2012.pdf (Accessed 17/03/2014).

UN-Water (2012a) *World Water Development Report 4: Managing Water Under Uncertainty and Risk* available at http://www.unesco.org/new/en/natural-sciences/environment/water/wwap/wwdr/wwdr4-2012/ (Accessed 17/03/2014).

UN-Water (2014) *A Post-2015 Global Goal for Water: Synthesis of Key Findings and Recommendations from UN-Water* available at http://water-1.iisd.org/news/un-water-synthesizes-recommendations-on-a-post-2015-global-goal-for-water/ (Accessed 27/03/2014).

UN-Water/UNDESA/UNICEF (2013) *The Post-2015 Water Thematic Final Consultation Report* available at http://www.worldwewant2015.org/water (Accessed 14/04/2014).

US Government Accountability Office (2013) *Making the Clean Water Act Work* available at http://www.gao.gov/assets/660/659496.pdf (Accessed 24/03/2014).

Walker, A. (2009) *The Independent Review of Charging for Household Water and Sewerage Services* available at https://www.gov.uk/government/uploads/system/uploads/attachment_data/file/69459/walker-review-final-report.pdf (Accessed 27/03/2014).

Water Resources (Scotland) Bill SP15 Policy Memorandum available at http://www.scottish.parliament.uk/parliamentarybusiness/Bills/52620.aspx (Accessed 24/03/2014).

Water Services (Intra-Group Regulation) Direction 2006 available at http://www.watercommission.co.uk/UserFiles/Documents/history/061031%20Section%2011%20Direction%20(Intra-Group).pdf (Accessed 26/03/2014).

Waterwise (2007) *Hidden Waters* available at http://www.waterwise.org.uk/resources.php (Accessed 03/04/2014).

Wentworth Group (2012) *Statement on the 2011 Draft Murray–Darling Basin Plan* available at http://wentworthgroup.org/wp-content/uploads/2013/10/Statement-on-the-2011-Draft-Basin-Plan.pdf (Accessed 20/03/2014).

WHO (2004) *Guidelines for Drinking Water Quality* 3rd Edition available at http://www.who.int/water_sanitation_health/dwq/gdwq3/en/ (Accessed 7/04/2007).

WHO (2006) *Guidelines for the Safe Use of Wastewater, Excreta and Greywater* available at http://www.who.int/water_sanitation_health/wastewater/gsuww/en/index.html (Accessed 27/03/2014).

WHO/UNICEF (2006) *Meeting the MDG Drinking Water and Sanitation Target* available at http://www.who.int/water_sanitation_health/monitoring/jmp2006/en/index.html (Accessed 27/03/2014).

WHO/UNICEF (2012) *Joint Monitoring Programme Progress on Drinking Water and Sanitation 2012 Update* available at http://www.who.int/water_sanitation_health/publications/2012/jmp_report/en/ (Accessed 17/03/2014).

WHO/UN-Water (2012) *Global Analysis and Assessment of Sanitation and Drinking-Water* available at http://www.un.org/waterforlifedecade/pdf/glaas_report_2012_eng.pdf.

WICS (2011) *Retail Competition in Scotland: An Audit Trail of the Costs Incurred and the Savings Achieved* available at http://www.watercommission.co.uk/UserFiles/Documents/WICSAuditTrail(B)%20(2).pdf (Accessed 26/03/2014).

WICS (2013) *Strategic Review of Charges 2015–21 Innovation and Choice* available at http://www.watercommission.co.uk/view_Price_Setting_2015-21.aspx (Accessed 27/03/2014).

Winpenny, J. (2003) *Report of the World Panel on Financing Water Infrastructure (Camdessus Report)* available at http://www.financingwaterforall.org/index.php?id=1098 (Accessed 27/03/2014).

World Bank (1992) *Governance and Development* available at http://elibrary.worldbank.org/doi/book/10.1596/0-8213-2094-7# (Accessed 16/04/2014).

World Bank (1993) *Water Resources Management: A World Bank Policy Paper* available at http://elibrary.worldbank.org/doi/book/10.1596/0-8213-2636-8 (Accessed 27/03/2014).

World Bank (2004) *Water Resources Sector Strategy* available at http://www-wds.worldbank.org/servlet/WDSContentServer/WDSP/IB/2004/06/01/000090341_20040601150257/Rendered/PDF/28114.pdf (Accessed 16/04/2014).

World Bank (2010) *Sustaining Water for All in a Changing Climate* (Progress Report) available at http://siteresources.worldbank.org/NEWS/Resources/sustainingwater.pdf (Accessed 16/04/2014).

World Bank/PPIAF (2006) *Water and Sanitation Privatization Toolkit* available at http://siteresources.worldbank.org/FINANCIALSECTOR/Resources/282044-1307652042357/Water_Full.pdf (Accessed 26/03/2014).

World Development Movement (2007) *Going Public: Southern Solutions for the Global Water Crisis* available at http://www.euractiv.com/en/sustainability/eu-water-policy-fire-developing-world/article-162544 (Accessed 01/04/2014).

Wouters, P. (2010) *Water Security: Global, Regional and Local Challenges* Institute for Public Policy Research.

WWF (2011) *Legal Action Pays Off* available at http://www.wwf.org.uk/wwf_articles.cfm?unewsid=4881 (Accessed 19/03/2014).

Young, M. (2010) *Environmental Effectiveness and Economic Efficiency of Water Use in Agriculture: The Experience of and Lessons from the Australian Water Reform Programme* OECD available at http://www.myoung.net.au/water/publications/OECD_Lessons_paper.pdf (Accessed 16/04/2014).

Websites

'Abbot Point Capital Dredging Project' available at http://www.gbrmpa.gov.au/about-us/consultation/current-proposals/abbot-point-capital-dredging-project (Accessed 24/03/2014).

'Ancient Greek Philosophy' available at http://www.iep.utm.edu/g/greekphi.htm (Accessed 24/03/2014).

'Australian Bureau of Statistics' available at http://www.abs.gov.au/ (Accessed 07/04/2014).

'Australian Government Bureau of Meteorology' available at http://www.bom.gov.au/climate/current/annual/qld/summary.shtml (Accessed 07/04/2014).

'Australian Government Geosciences Australia' available at http://www.ga.gov.au/education/geoscience-basics/dimensions/area-of-australia-states-and-territories.html (Accessed 07/04/2014).

'Australian Government Marine Protection' available at http://www.environment.gov.au/coasts/index.html (Accessed 21/03/2014).

'Australian Network of Environmental Defenders' Offices' available at http://www.edo.org.au/ (Accessed 24/03/2014).

'Better Regulation' available at https://www.gov.uk/government/organisations/better-regulation-delivery-office (Accessed 27/03/2014).

'Bulk Water Prices' available at http://www.dews.qld.gov.au/policies-initiatives/water-sector-reform/water-pricing/bulk-water-prices (Accessed 27/03/2014).

'Caring for our Country' available at http://www.nrm.gov.au/index.html (Accessed 20/03/2014).

'Catchment Based Approach' available at http://www.environment-agency.gov.uk/research/planning/131506.aspx (Accessed 26/03/2014).

'Catchment Sensitive Farming' available at http://www.naturalengland.org.uk/ourwork/farming/csf/default.aspx (Accessed 24/03/2014).

'Coastal Management in Queensland' available at http://www.ehp.qld.gov.au/coastal/management/ (Accessed 21/03/2014).

'Condamine Catchment Management Association' available at http://www.condaminecatchment.com.au/ (Accessed 20/03/2014).

'Customer Engagement' available at http://www.ofwat.gov.uk/pricereview/pr14/customer/ (Accessed 27/03/2014).

'Customer Forum' available at http://customerforum.org.uk/ (Accessed 27/03/2014).

'Development of the Murray–Darling Basin Plan' documents available at http://www.mdba.gov.au/what-we-do/basin-plan/development/steps-in-the-development-of-the-basin-plan (Accessed 20/03/2014).

'Drinking Water Quality Management Plans' available at http://www.dews.qld.gov.au/water-supply-regulations/drinking-water/drinking-water-quality-management-plans (Accessed 27/03/2014).

'Dwr Cymru Welsh Water' available at http://www.dwrcymru.com/en/Company-Information.aspx (Accessed 26/03/2014).

'Environment Agency Abstraction Charges Scheme 2013/14' available at http://www.environment-agency.gov.uk/business/regulation/38809.aspx (Accessed 24/03/2014).

'Environment Agency Drought Planning' available at http://www.environment-agency.gov.uk/homeandleisure/drought/31771.aspx (Accessed 19/03/2014).

'Environment Agency National Liaison Panel' membership and papers available at http://www.environment-agency.gov.uk/research/planning/33114.aspx (Accessed 19/03/2014).

'Farming Grants, Subsidies and Services' available at http://www.scotland.gov.uk/Topics/farmingrural/Agriculture/grants (Accessed 24/03/2014).

'French Civil Code' available at http://www.lexinter.net/ENGLISH/civil_code.htm (Accessed 24/03/2014).

'Global Water Partnership' available at http://www.gwp.org/ (Accessed 17/03/2014).

'Goal 7 Environmental Sustainability' available at http://www.un.org/millenniumgoals/environ.shtml (Accessed 27/03/2014).

'Government of South Africa: About South Africa' available at http://www.gov.za/aboutsa/geography.htm (Accessed 07/04/2014).

'Great Barrier Reef Strategic Assessment' available at http://www.dsdip.qld.gov.au/gbr-strategic-assessment/ (Accessed 24/03/2014).

'GWP ToolBox' available at http://www.gwptoolbox.org/ (Accessed 19/03/2014).

'Healthy Waterways Partnership' available at http://www.healthywaterways.org/AboutUs/PartnershipOffice.aspx (Accessed 20/03/2014).

'Hydrology, Environment, Life and Policy' available at http://www.unesco.org/new/en/natural-sciences/environment/water/ihp/ihp-programmes/help/ (Accessed 31/03/2014).

'Instruments of Appointment' available at http://www.ofwat.gov.uk/industry structure/licences/ (Accessed 27/03/2014).

'Interim Determinations' available at http://www.ofwat.gov.uk/pricereview/setting/interim/ (Accessed 27/03/2014).

'Internet Encyclopedia of Philosophy' available at http://www.iep.utm.edu/g/greekphi.htm (Accessed 17/03/2014).

'June Returns' available at http://www.ofwat.gov.uk/regulating/junereturn/ (Accessed 27/03/2014).

'Key Indicators Guidance' available at https://www.ofwat.gov.uk/regulating/compliance/reportingperformance/kpi (Accessed 27/03/2014).

'LandCare' and 'Natural Resource Management' Programmes available at http://www.daff.gov.za/daffweb3/Programmes (Accessed 24/03/2014).

'Local Management Arrangements for Irrigation Channel Schemes' available at http://www.lmairrigation.com.au/ (Accessed 24/03/2014).

'MadSci Network' available at http://www.madsci.org/posts/archives/may2000/958588306.An.r.html (Accessed 24/03/2014).

'Metropolitan Glasgow Strategic Drainage Partnership' see generally http://www.mgsdp.org/ (Accessed 27/03/2014).

'Millennium Development Goal 7' available at http://www.un.org/millenniumgoals/environ.shtml (Accessed 15/04/2014).

'Millennium Ecosystem Assessment' (2005) available at http://www.maweb.org/en/Index.aspx (Accessed 25/03/2014).

'Murray–Darling Basin Plan Consultation Documents' available at http://www.mdba.gov.au/what-we-do/basin-plan/consultation (Accessed 20/03/2014).

'Murray–Darling Basin' available at http://www.mdba.gov.au/ (Accessed 20/03/2014).

'National Water Quality Management Strategy' available at http://www.environment.gov.au/water/policy-programs/nwqms/index.html (Accessed 20/03/2014).

'OFWAT Publications and Reports' available at http://www.ofwat.gov.uk/publications#reports (Accessed 27/03/2014).

'Population Estimates Scotland 2011–12' available at http://www.gro-scotland.gov.uk/files2/stats/population-estimates/mid2012/j29078400.htm (Accessed 07/04/2014).

'Pricing and Economic Regulation Reforms Project' available at http://www.dwaf.gov.za/Projects/PERR/Default.aspx (Accessed 27/03/2014).

'Private Water Supplies' available at http://dwi.defra.gov.uk/stakeholders/private-water-supplies/index.htm (Accessed 26/03/2014).

'Public Sector International Research Unit' University of Greenwich available at www.psiru.org/ (Accessed 01/04/2014).

'Queensland Bulk Water Prices' available at http://www.dews.qld.gov.au/policies-initiatives/water-sector-reform/bulk-water-prices (Accessed 24/03/2014).

'Queensland Government Statistician's Office' available at http://www.oesr.qld.gov.au/products/briefs/pop-growth-qld/qld-pop-counter.php (Accessed 07/04/2014).

'Queensland Government Sustainable Agriculture' available at http://www.daff.qld.gov.au/environment/sustainable-agriculture (Accessed 24/03/2014).

'Queensland Government Water Quality and Water Use' available at http://www.daff.qld.gov.au/environment/sustainable-agriculture/water-quality-and-water-use (Accessed 24/03/2014).

'Queensland Water Resource Plans' available at http://www.nrm.qld.gov.au/wrp/ (Accessed 20/03/2014).

'REACH: Registration, Evaluation, Authorisation and Restriction of Chemicals' available at http://ec.europa.eu/enterprise/sectors/chemicals/reach/index_en.htm (Accessed 26/03/2014).

'SCaMP Project' available at http://www.rspb.org.uk/ourwork/projects/details/218780-scamp-sustainable-catchment-management-programme (Accessed 26/03/2014).

'Scotland Info' available at http://www.scotlandinfo.eu/weather-climate.html (Accessed 07/04/2014).

'SEPA Area Advisory Groups Membership and Papers' available at http://www.sepa.org.uk/water/river_basin_planning/area_advisory_groups.aspx (Accessed 19/03/2014).

'SEPA National Advisory Group Membership and Papers' available at http://www.sepa.org.uk/water/river_basin_planning/national_advisory_group.aspx (Accessed 19/03/2014).

'SEQ Catchments' available at http://www.seqcatchments.com.au/ (Accessed 20/03/2014).

'South Africa Water Resources Council' available at http://www.wrc.org.za/Pages/Resources_Regionalstats.aspx (Accessed 03/04/2014).

'South Africa's Interaction with SADC' available at http://www.dfa.gov.za/foreign/Multilateral/africa/sadc.htm (Accessed 19/03/2014).

'Steps in the Development of the Basin Plan' available at http://www.mdba. gov.au/what-we-do/basin-plan/development/steps-in-the-development-of-the-basin-plan (Accessed 15/04/2014).

'UK Government Office of National Statistics' available at http://www.ons. gov.uk/ons/taxonomy/index.html?nscl=Population (Accessed 07/04/2014).

'UK Meteorological Office Climate and Rainfall' available at http://www. metoffice.gov.uk/climate/uk/actualmonthly/ (Accessed 07/04/2014).

'UKTAG WFD' available at http://www.wfduk.org/ (Accessed 26/03/2014).

'Upstream Thinking' available at http://www.wrt.org.uk/ (Accessed 26/03/ 2014).

'Water Authorities' available at http://www.nrm.qld.gov.au/water/regulation/ water_authorities (Accessed 27/03/2014).

'Water Environment (Controlled Activities) Fees And Charges (Scotland) Scheme 2013' available at http://www.sepa.org.uk/about_us/charging_ schemes/current_charging_schemes.aspx (Accessed 24/03/2014).

'Water Publications' available at http://www.ehp.qld.gov.au/water/monitoring/ publications.html (Accessed 24/03/2014).

'Water Resource Planning' available at http://www.dnrm.qld.gov.au/water/ catchments-planning/qmdb/water-resource-planning (Accessed 24/03/2014).

'Water Supply Licences' available at https://www.ofwat.gov.uk/competition/ wsl/wsllicensees/ (Accessed 26/03/2014).

'Water Tribunal Case Decisions' available at http://www.dwaf.gov.za/Water Tribunal/Cases.aspx (Accessed 24/03/2014).

'WICS Publications' available at http://www.watercommission.co.uk/view_ Publications_Main.aspx (Accessed 27/03/2014).

'Wivenhoe Dam Class Action' available at https://www.imf.com.au/wiven hoe/ (Accessed 20/03/2014).

'Working For Water' and 'Working for Wetlands' Programmes available at https://www.environment.gov.za/projectsprogrammes (Accessed 24/03/ 2014).

'World Bank Annual Freshwater Withdrawals' available at http://data.world bank.org/indicator/ER.H2O.FWAG.ZS/countries (Accessed 03/04/2014).

'World Bank Renewable Internal Water Resources' available at http://data. worldbank.org/indicator/ER.H2O.INTR.PC (Accessed 07/04/2014).

'World Water Council' available at http://www.worldwatercouncil.org/index. php?id=1 (Accessed 17/03/2014).

'WQM in SA' available at http://www.dwaf.gov.za/Dir_WQM/wqmFrame. htm (Accessed 24/03/2014).

Index

Printed in the United States
By Bookmasters